T-Labs Series in Telecommunication Services

Series editors

Sebastian Möller, Berlin, Germany
Axel Küpper, Berlin, Germany
Alexander Raake, Berlin, Germany

More information about this series at http://www.springer.com/series/10013

Marc Halbrügge

Predicting User Performance and Errors

Automated Usability Evaluation Through
Computational Introspection of Model-Based
User Interfaces

 Springer

Marc Halbrügge
Quality and Usability Lab
TU Berlin
Berlin
Germany

ISSN 2192-2810 ISSN 2192-2829 (electronic)
T-Labs Series in Telecommunication Services
ISBN 978-3-319-86849-3 ISBN 978-3-319-60369-8 (eBook)
DOI 10.1007/978-3-319-60369-8

Printed on acid-free paper

This Springer imprint is published by Springer Nature
The registered company is Springer International Publishing AG
The registered company address is: Gewerbestrasse 11, 6330 Cham, Switzerland

Contents

Acronyms

ACT-R	Adaptive Control of Thought–Rational (Anderson et al. 2004)
ANOVA	ANalysis Of VAriance
AOI	Area Of Interest (eye-tracking)
AUE	Automated Usability Evaluation
AUI	Abstract User Interface (model that is part of the CAMELEON reference framework)
CAMELEON	Context Aware Modelling for Enabling and Leveraging Effective interactiON (Calvary et al. 2003)
CI	Confidence Interval
CREAM	Cognitive Reliability and Error Analysis Method (Hollnagel 1998)
CTT	ConcurTaskTree (Paternò 2003)
CUI	Concrete User Interface (model that is part of the CAMELEON reference framework)
DBDR	Display-Based Difference Reduction (Gray 2000)
ETA	Embodied cognition-Task-Artifact triad (Gray 2000)
FUI	Final User Interface (as part of the CAMELEON reference framework)
GLEAN	GOMS Language Evaluation and ANalysis (Kieras et al. 1995)
GLMM	Generalized Linear Mixed Model
GOMS	Goals, Operators, Methods, and Selection Rules (Card et al. 1983)
GUI	Graphical User Interface
GUM	Generic model of cognitively plausible user behavior (Butterworth et al. 2000)
HCI	Human–Computer Interaction
HMM	Hidden Markov Model
HTA	Hierarchical Task Analysis
HTML	HyperText Markup Language (Raggett et al. 1999)
ISO	International Organization for Standardization
KLM	Keystroke-Level Model (Card et al. 1983)
LMM	Linear Mixed Model

LTMC	Long-Term Memory/Casimir (Schultheis et al. 2006)
LTM	Long-Term Memory
MANOVA	Multivariate ANalysis Of VAriance
MASP	Multi-Access Service Platform (Blumendorf et al. 2010)
MBUID	Model-Based User Interface Development (Meixner et al. 2011)
MFG	Memory for Goals (Altmann and Trafton 2002)
MHP	Model Human Processor (Card et al. 1983)
MLSD	Maximum Likely Scaled Difference (Stewart and West 2010)
PC	Personal Computer
RMSE	Root Mean Squared Error
TCT	Task Completion Time
TERESA	Transformation Environment for inteRactivE Systems representAtions (Mori et al. 2004)
UCD	User-Centered Design (Gould and Lewis 1985)
UI	User Interface
WMU	Working Memory Updating (Ecker et al. 2010)
WYSIWYG	What You See Is What You Get
XML	eXtensible Markup Language (Bray et al. 1998)

List of Figures

List of Tables

Chapter 1
Introduction

In this chapter:

- What is usability, why is it important?
- The dilemma of maintaining usability for multi-target systems
- How model-based development help creating multi-target systems
- Research direction: Can the model-based approach help to predict the usability of such systems as well?

The main insight in the field of human-computer interaction (HCI) is that application systems must not only function as specified, they must also be usable by humans. What does that mean?

1.1 Usability

As defined in ISO 9241-11 (1998), the usability of a system can be decomposed into three aspects:

- Effectiveness (the accuracy and completeness with which users achieve specified goals)
- Efficiency (the resources expended in relation to the accuracy and completeness with which the users achieve goals)
- Satisfaction (the comfort and acceptability of use)

These aspects do overlap to some degree (e.g., low effectiveness may lead to lower efficiency in the presence of additional corrective actions), but are still sufficiently different from each other. While effectiveness and efficiency can be measured objectively (e.g., task completion time, task success rate), user satisfaction needs subjective

© Springer International Publishing AG 2018
M. Halbrügge, *Predicting User Performance and Errors*, T-Labs Series
in Telecommunication Services, DOI 10.1007/978-3-319-60369-8_1

measurement (e.g., questionnaires). Compared to the other two aspects, satisfaction is more broadly defined and multi-faceted. In addition to the already mentioned comfort and acceptability, it may also comprise notions of aesthetics and identification with the product (e.g., Hassenzahl et al. 2015).

Why is Usability Important?
On the side of the user (or customer), bad usability in terms of effectiveness and efficiency first of all leads to low productivity. This stretches from negligible delayed task completion to severe consequences in safety-critical environments (e.g., medical, machine control, air traffic) if bad usability leads to operator errors.[1] Bad usability in terms of satisfaction may lead to low enjoyment of use (Hassenzahl et al. 2001) which in consequence may lead to decreased frequency of use.

On the side of the supplier of the application or product, this may in turn lead to increased support costs (e.g., if the customers cannot attain their goals due to usability problems; Bevan 2009) and decreased product success (Mayhew 1999). As a consequence, the revenue of the supplier may be at risk, and/or increased development spending may occur if unplanned usability updates of the application are necessary (Bevan 2009).

> In summary: Usability is not an optional feature, it is a prerequisite of the success of a product given a fixed amount of development time and cost.

Usability Engineering
In order to achieve usable systems, the principles of User-Centered Design (Gould and Lewis 1985; ISO 9241-210 2010) should guide the development of an application:

- early focus on users and tasks
- empirical measurement
- iterative design

Nielsen (1993) builds on these principles in his model of the Usability Engineering Lifecycle which ties them more closely to the different stages of product development (e.g., initial analysis, roll-out to the customer).

The details of Nielsen's model are beyond the scope of this work, but the methods to attain principles of user-centered design will be important in the following. In order to focus on the users' tasks, methods in the broad field of *task analysis* (Kirwan and Ainsworth 1992) are applied. This means building (often hierarchical) models of the actions that users take to attain their goals. These model can then be used to guide the design of the user interface.

---

[1] See Reason (2016) for examples like commonly designed drop-down lists leading to false medical prescription leading to overdose and patient death.

Empirical measurement is mainly achieved through *user tests* (Nielsen and Landauer 1993) where actual users are observed while performing tasks with the application or a mock-up thereof. User tests with small samples (e.g., $N = 8$) are already very successful in eliciting usability problems like misleading element captions or bad choice of fonts and button sizes. Larger samples allow additional measurements like user satisfaction questionnaires or deeper error analyses.

1.2 Multi-Target Applications

The models and processes that have been formulated in the 1980s and 1990s are still valid today, but are facing increasing difficulties with the recent explosion[2] of the number of mobile appliances and device types.

With regards to software development and UI design, this leads to several problems. Applications must now work not just on one, but on several kinds of devices with different form factors (e.g., traditional PC, tablet, smart phone, smart television) and interaction paradigms (e.g., point and click, touch, voice, gesture). In principle, this could be approached by reimplementing an application for every target system, but this would lead to massively increased development and maintenance costs. A better solution is to create a single *multi-target* application which allows easy adaption of the UI to different form factors and interaction paradigms. Developing a multi-target application results in higher development costs than developing a single-target application, but should reduce costs compared to maintaining several target-specific versions of an application at the same time.

Besides the *engineering* challenge posed by multi-target applications (how to develop efficiently), maintaining their *usability* is an equally challenging task. If an application supports a multitude of device targets, its usability has to be ensured on any of these. But the methods of user-centered design, esp. the aspect of empirical measurement, do not scale well. Running user tests on many different devices would be extremely costly and time consuming. A possible alleviation of this situation that will be further elaborated in the following is Automated Usability Evaluation (AUE; Ivory and Hearst 2001, details in Chap. 4).

On the engineering side of the problem, a promising solution for keeping the development costs of multi-target applications at bay is the process of Model-Based UI Development (MBUID; Calvary et al. 2003, details in Chap. 3). In the context of this work, MBUID has an interesting aspect: If applied using model-based runtime frameworks (definition in Sect. 3.1), this development process does not create a monolithic application at the end, but allows to enumerate the current elements of its UI through computational *introspection* with additional pointers to

---

[2]Numbers for Germany to justify the use of the word "explosion": In 2010, there were 80.7 stationary and 57.8 mobile personal computers (PC) per 100 households. In 2015, the number of mobile computers has more than doubled (133.2 per 100 households, 39.1 thereof being tablets) while the number of stationary computers slightly declined to 63.1 (Statistisches Bundesamt 2016).

computer-processable meta-information, e.g., task hierarchies as result of an initial task analysis. As this closely resembles the methods of user-centered design given above, this information could be useful to create dedicated tools for the automated usability evaluation of model-based applications. How could this be achieved?

1.3   Automated Usability Evaluation of Model-Based Applications

In order to utilize the additional information provided by model-based applications for predicting their usability, a link between the properties of the model-based UI on the one hand and different aspects of usability on the other hand must be established.

This link (or 'function') must have two important properties. First, it must be valid, i.e., its predictions must resemble human behavior as it can be observed during user tests as closely as possible. In this work, validity will be ensured by basing the link function on empirical results and psychological theory.

Second, the link between introspectable properties of the UI and its usability must be suitable for automation. This excludes all techniques which rely on the application of heuristics or vague principles by human analysts. The method of choice to achieve automatibility is computational *cognitive modeling* (Byrne 2013). The application of cognitive models that implement psychological theory is also a means to ensure the validity of the theoretical assumptions as it may elicit gaps in the theory and forces to exemplify the theory to the extent that it actually becomes implementable as software (e.g., the notion of "x increases with y" has to become "$x = -2 + y^3$").

1.4   Research Direction

Having set the domain and overall goal of this work, an initial research question can be stated. This shall guide the presentation of theoretic accounts and related work in the following chapters. A refined research question will be given at the end of the next part.

> Research direction: How can UI meta-information as created by the MBUID process be used for automated usability evaluation?

Further questions can be derived from this, e.g., which parts of usability can be predicted based on this meta-information? How well can they be predicted? To which extent can this be automated, how much human intervention is necessary?

1.5 Conclusion

Maintaining the usability of multi-target applications is a daunting task. It might be alleviated if automated usability predictions were available from early stages of development on. The preparation and validity of such predictions could be facilitated and improved through incorporation of meta-information of the applications that is available if their development follows the MBUID process. The goal of this work is to analyze whether this last proposition actually holds.

This question is approached the following way: The next part will provide the necessary fundamentals on psychological theory needed to create predictions of usability, the nature of multi-target MBUID applications, and existing solutions for automated usability evaluation. At the end of the part, a refinement of the broad research question given above will be possible.

The following part gives the empirical groundwork and derives psychologically plausible models of how the efficiency and effectiveness of the UI of a specific multi-target MBUID application are determined by properties of its UI design.

An actual implementation of an error prediction system for MBUID applications based on these psychological models and the meta-information provided by the model-based application framework is presented in the third and last part alongside its validation on a different application. This is followed by an analysis of the automatic predictability of the remaining third aspect of usability (user satisfaction), a general discussion of the strengths and limitations of the approach, and final concluding remarks.

Part I
Theoretical Background
and Related Work

Chapter 2
Interactive Behavior and Human Error

In this chapter[1]:

- Usability is about how users use systems, i.e., user *behavior*. How is this characterized? What drives it?
- Major properties of user behavior regarded here are a) the time needed and b) the errors made. What distinguishes erroneous from 'normal' behavior?
- Which types of errors are important in HCI and how can these be explained theoretically?

The basic assumption of this work is that the behavior of a user of a system depends largely on the interface of the system. As John and Kieras have stated:

> Human activity with a computer system can be viewed as executing methods to accomplish goals, and because humans strive to be efficient, these methods are heavily determined by the design of the computer system. This means that the user's activity can be predicted to a great extent from the system design. (John and Kieras 1996a)

In other words: Given a sufficiently detailed definition of the user interface, one should be able to predict user behavior. What exactly follows from John and Kieras' proposition that "humans strive to be efficient" may be arguable, though. There is usually a tradeoff between effort and time. And the 'sweet spot' that gives the best result may differ between people and contexts.[2]

Starting from the premise given above, the factors that shape interactive behavior can be stated more formally. One such formalism is the ETA-triad (Gray 2000) as shown in Fig. 2.1. Understanding interactive behavior depends on understanding in

---

[1]Parts of Sect. 2.3 have already been published in Halbrügge et al. (2015b).

[2]Example: Keying ahead without visual feedback can save time, but needs more cognitive resources than pure reaction to visual cues on the interface.

© Springer International Publishing AG 2018
M. Halbrügge, *Predicting User Performance and Errors*, T-Labs Series
in Telecommunication Services, DOI 10.1007/978-3-319-60369-8_2

Fig. 2.1 The ETA-triad
(Gray 2000)

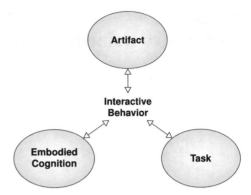

three areas: the task, the artifact, and the constraints of the perception-cognition-motor system of the embodied[3] user.

In the context of this work, the schema of the ETA-triad is first applied to predict both successful and unsuccessful user behavior. These predictions shall then form the basis of automated usability predictions of the 'artifact' part of the triad.

The current chapter describes the basic assumptions and models used to describe the properties of the embodied user. It starts with a cognitive engineering model of human action control and error, followed by a phenotypical classification system of human error. This is finally contrasted with a theory-driven attempt to explain why errors happen while pursuing goals with a computer system.

2.1 Action Regulation and Human Error

A sufficiently detailed model of human decision-making for cognitive engineering domain has been proposed by Rasmussen (1986). His *step-ladder model* (see Fig. 2.2) consists of a perceptual leg on the left and an action leg on the right. The decision making process is started by activation through some new percept. This may trigger an attentional shift ("Observe") and conscious processing of the percept ("Identify"). Interpretation and evaluation leads to the definition of a new goal ("Define Task") and/or trigger an already existing action sequence ("Stored Procedure") which are finally executed.

Most activities in daily life do not require to go up to the very top of the ladder. Shortcuts between any two elements of the ladder may be acquired through learning. Examples for such shortcuts are given in Fig. 2.2.

---

[3]In this work, the term "embodiment" is used in a more elaborated sense than "cognition with added perception and motor capabilities". Instead, "embodied cognition" means that the analysis is not targeting the mind of the user, but the user-artifact dyad. In terms of Wilson's six views of embodied cognition, this is mainly related to the aspects "We off-load cognitive work onto the environment" and "The environment is part of the cognitive system" (Wilson 2002).

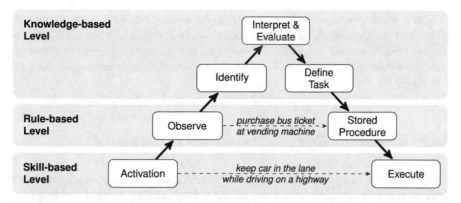

Fig. 2.2 Step-ladder model of decision making. Adapted and simplified from Rasmussen (1986, p. 67), Hollnagel (1998, p. 61), Reason (1990, p. 64), and Rasmussen (1983). *Solid arrows* display the expected sequence of processing stages during decision-making or problem solving. *Dashed arrows* represent examples for shortcuts that have been established mainly through training. Such shortcuts may exist between any two of the boxes

Another view on the step-ladder model is the level of action control that is applied. These levels are represented as gray boxes in the background of the figure. According to Rasmussen (1983), human action control can be described on three levels: skill-, rule-, and knowledge-based behavior. Skill-based behavior on Rasmussen's lowest level is generated from highly automated sensory-motor actions without conscious control. Knowledge-based behavior on the other hand is characterized by explicit planning and problem solving in unknown environments. In between the skill and the knowledge levels is rule-based behavior. While being goal-directed, rule-based actions do not need explicit planning or conscious processing of the current goal. The stream of actions follows stored procedures and rules that have been developed during earlier encounters or through instruction. Interaction with computer systems is mainly located on this intermediate rule-based level of action control.

2.1.1   Human Error in General

Human error is commonly defined as

> those occasions in which a planned sequence of mental or physical activities fail to achieve its intended outcome, [and] when these failures cannot be attributed to the intervention of some chance agency. (Reason 1990, p. 9)

This definition is very broad as it is meant to cover any kind of erroneous action. It is nevertheless instrumental in highlighting a property of human error that makes its research so complicated: Whether something is an error or not depends on its *intended outcome*. This has two consequences. First, it is impossible to determine whether something is an error or not without knowing the (unobservable) intention behind it. And second, errors usually cannot be observed as they happen, but only when their outcome becomes manifest. In terms of the step-ladder model, this means that only the "Execute" stage on the lower right creates overt behavior. In case of an error, the cause of the error may be located on any other stage or connection between two stages.

2.1.2 Procedural Error, Intrusions and Omissions

Interaction with computer systems is mainly located on the intermediate rule-based level of action control. On this level, behavior is generated using stored rules and procedures that have been formed during training or earlier encounters. Errors on the rule-based level are not very frequent (below 5%), but pervasive and cannot be eliminated through training (Reason 1990). While Norman (1988) subsumes these errors within the 'slips' category, Reason (1990, 2016) refers to them as either 'lapses' in case of forgetting an intended action or 'rule-based mistakes' when the wrong rule (i.e., stored procedure) is applied. Because of this ambiguity, the term *procedural error* is used throughout this work.

Procedural error is defined as the violation of the (optimal) path to the current goal by a non-optimal action (cf., Singleton 1973). This can either be the addition of an unnecessary or even hindering action, which is called an *intrusion*. Or a necessary step can be left out, constituting an *omission*.

How does procedural error manifest itself in daily life?

Example: Postcompletion Error
A very common and also well researched example of procedural error is *postcompletion error* (Byrne and Bovair 1997; Byrne and Davis 2006). This type of error is characterized by the omission of the last step of an action sequence if the overall goal of the user has already been accomplished before. Typical examples of postcompletion errors are forgetting the originals in a copy machine[4] and leaving the bank card in a teller machine after having taken the money.

Similar errors can happen during the first step of an action sequence as well. This type has been coined *initialization error* (Li et al. 2008; Hiltz et al. 2010). An example of this kind of error is forgetting to press a 'mode' key before setting the alarm clock on a digital watch (Ament et al. 2010).

Because of its prototypical nature, postcompletion error has become one of the basic tests for error research and action control theories. The ability of such theories to explain postcompletion error is often used as argument in favor of their validity (e.g., Byrne and Davis 2006; Altmann and Trafton 2002; Butterworth et al. 2000, see Sect. 2.3 below). Before reviewing theories of procedural error, its notion shall first be contrasted with other descriptions of typical errors that occur at home and in the workplace.

---

[4]Side remark: According to a copy shop clerk, this error has been superseeded in frequency by clients forgetting their data stick after having received their printout.

2.2   Error Classification and Human Reliability

Early error research has focused on classification systems of human error. These contain usually more categories than the differentiation between omissions and intrusions given above. The best known of these classification systems has been created by Norman and will be presented in the following.

2.2.1   Slips and Mistakes—The Work of Donald A. Norman

Norman (1981, 1988) distinguished between slips and mistakes. The basic difference between those is that mistakes happen when an incorrect intention is acted out correctly. Slips on the other hand mark situations when a correct intention is acted out incorrectly. Referring to the step-ladder model above (Fig. 2.2), mistakes belong to knowledge-based behavior in the upper part of the ladder and slips belong to either rule-based or skill-based behavior. Norman (1988) does not provide further sub-categories for mistakes, but distinguishes between several types of action slips. These are given with examples in Table 2.1.

Norman's classification scheme has drawn criticism for several reasons. First, according to Hollnagel (1998), it mingles genotypes (e.g., 'associative activation error') and phenotypes (e.g., 'mode error') which leads to inconsistencies. And second, it is disputable whether mode errors are actual action slips in Norman's terms as they are not characterized by faulty action. A 'mistaken system state' should rather be considered an incorrect intention, which puts mode errors into the 'mistake' category.

In the context of this work, of highest importance is whether such a classification suits automatable usability predictions. How does Norman's system relate to these?

2.2.2   Human Reliability Analysis

Classification schemes like Norman's have been combined to models of *human reliability* that can be used to predict overall error rates for safety-related tasks like controlling a nuclear power plant (Kirwan 1997a, b). Unfortunately, the validity of these approaches does not live up to the expectations (Wickens et al. 2015, Chap. 9). This is most probably due to the fact that classification schemes only *describe* human error, but do not *explain* how correct and erroneous behavior is produced. The remainder of this chapter presents models of human behavior and action control that try to provide such explanations.

Table 2.1 Types of action slips and examples as reported in Norman (1988, p. 107f). The examples have been slightly shortened by the author

Type of slip	Example
Capture error (habit take-over)	"I was using a copying machine, and I was counting the pages. I found myself counting '1, 2, 3, 4, 5, 6, 7, 8, 9, 10, Jack, Queen, King.' I have been playing cards recently."
Description error (incomplete specification)	"A former student reported that one day he came home from jogging, took off his sweaty shirt, and rolled it up in a ball, intending to throw it in the laundry basket. Instead he threw it in the toilet. (It wasn't poor aim: the laundry basket and the toilet were in different rooms.)"
Data driven error (dominant stimulus driven)	"I was assigning a visitor a room to use. I decided to call the department secretary to tell her the room number. I used the telephone with the room number in sight. Instead of dialing the secretary's phone number I dialed the room number."
Associative activation error (Freudian slip)	"My office phone rang. I picked up the receiver and bellowed 'Come in' at it."
Loss-of-action error (forget intention)	"I have to go to the bedroom. As I am walking, I have no idea why I'm going there. I keep going hoping that something in the bedroom will remind me. Finally I realize that my glasses are dirty. I get my handkerchief from the bedroom, and wipe my glasses clean."
Mode error (mistaken system state)	"I had just completed a long run in what I was convinced would be record time. It was dark, so I could not read the time on my stopwatch. I remembered that my watch had a built-in light. I depressed the button, only to read a time of zero seconds. I had forgotten that in stopwatch mode, the same button cleared the time and reset the stopwatch."

2.3 Theoretical Explanations of Human Error

2.3.1 Contention Scheduling and the Supervisory System

Norman and Shallice (1986) proposed a model of action selection called 'contention scheduling' which depends on activation through either sensory (horizontal) 'triggers' or internal (vertical) 'source' schemas which represent volitional control by a so-called 'supervisory attentional system'. The 'contention' in this model arises from reciprocal inhibition of the schemas[5] that belong to individual actions. These rather simple assumptions can already explain some types of errors, e.g., capture errors as activation from a sensory trigger that overrides the action sequence that had been followed before. From the description of the model, it should already be clear that it does not cover errors on the knowledge-based or skill-based levels, but aims at routine activities like making coffee.

---

[5]Note: These are 'action' schemas, not to be confused with the 'source' schemas.

The contention scheduling model has been implemented by Cooper and Shallice (2000) with a subsequent validation using data from patient groups with impaired action control. Interestingly, they do not stick to the error categories that had be put forward by Norman (1988), but use a coding system based on the *disorganization* of actions—as opposed to errors—within a sequence instead (Schwartz et al. 1998). They write about Norman's classification system:

> These categories are neither disjoint nor definitive, and there can be difficulties in using them to classify certain action sequences (Cooper and Shallice 2000, p. 300)

Later research aimed at confirming the existence of a supervisory system based on action latencies while learning new routines (Ruh et al. 2010).

2.3.2 Modeling Human Error with ACT-R

A more rigorous attempt to explain human error based on psychological theory of action control has been presented by Gray (2000). He observed users while programming a videocassette recorder (VCR) to record television shows and modeled their behavior using the now outdated version 2 of the cognitive architecture ACT-R (Anderson and Lebiere 1998, see Sect. 4.2). According to Gray, using a cognitive model leads not only to better understanding of human error, it also creates better vocabulary for the description of errors than simple category systems like the one of Norman (1981, see Table 2.1).

Gray assumes that goals and subgoals that control behavior are represented in a hierarchical tree-like structure. The goal stack of ACT-R 2 is used to traverse this tree in a depth-first manner to produce actual behavior. In order to complete a goal, its first subgoal is pushed to the stack. After completion of the subgoal, it is popped from the stack and the next subgoal is pushed. Based on this process, errors can be divided into push errors (attaining a subgoal at an unpredicted point in time) and pop errors (releasing a subgoal too early or too late).

Push errors observed in Gray's VCR paradigm were for example setting 'rec-mode' before start and end time had been set (rule hierarchy failure) or trying to access something that is currently visible, but unchangeable (display-induced error). Push errors tend to decrease with practice (through learning of the goal hierarchy).

Pop errors can be further decomposed into premature pops (goal suspensions) and postponed pops. Premature pops manifest themselves by a subgoal being interrupted before it is completed (intrusion). Interrupting goals are often close to the interrupted ones. Interestingly, premature pops *increase* with routine. Gray attributes this to competition with leftovers from previous trials. Postponed pops on the other hand were mainly physical slips, e.g., too many repetitions while setting the clock to the start or end time of the show to be recorded.

Gray's cognitive model proved correct in the sense that it a) works, i.e., can solve the task, b) matches human behavior on correct trials, and c) makes similar errors that humans make. At the same time, the vision based strategy applied in the model serves as error detection and error recovery strategy as well. This is also in line with the error recovery behavior observed in the VCR paradigm. Of 28 detected and corrected errors, only four were not visible to the user. Gray concludes that error detection is local. Errors are detected and corrected either right after they have been made, or not at all.

Postcompletion errors (see Sect. 2.1.2) could be classified as premature pops in Gray's nomenclature, but Gray's model has problems explaining these. As ACT-R 2's goal stack has perfect memory, the model does not exhibit premature pops if there is no other goal that can take over control. At the end of an action sequence, no such intruder is available.

The approach taken by Gray provides important insights about how errors can be explained and showcases the usefulness of cognitive modeling as a research method in this field. The assumption of a goal hierarchy that is processed recursively using a stack has been questioned, though. An alternative, more parsimonious theory of action control is the Memory for Goals model (Altmann and Trafton 2002).

2.3.3   Memory for Goals Model of Sequential Action

The Memory for Goals model (MFG; Altmann and Trafton 2002) postulates that goals and subgoals are not managed using a dedicated goal stack, but reside in generic declarative memory. This implies that goals not 'special', but are memory traces among many others. As such, they are subject to the general characteristics of human associative memory (Anderson and Bower 2014), in particular time-dependent and noisy *activation*, *interference*, and *priming*. With respect to action control and human error, lack of activation of a subgoal can cause omissions, while interference with other subgoals can result in intrusions.

Based on these assumptions, postcompletion error (see Sect. 2.1.2) is mainly explained by lack of activation through priming. In the MFG, a sequence of actions arises from consecutive retrievals of subgoals from declarative memory. These retrievals are facilitated by priming from internal and external cues. As the subgoals that correspond to typical postcompletion errors (e.g., taking the originals from a copy machine) are only weakly connected to the overall goal of the action sequence (e.g., making copies), they receive less priming and are therefore harder to retrieve.

While the MFG theory initially has been validated on the basis of Tower-of-Hanoi experiments, i.e., rather artificial problem-solving tasks in the laboratory, it has been shown to generalize well to procedural error during software use and has been extensively used in the human-computer interaction domain (e.g., Li et al. 2008; Trafton et al. 2011).

2.4 Conclusion

The current chapter introduced the basic concepts related to human behavior with regards to cognitive engineering of interactive software systems. User behavior can be decomposed based on how control is applied. This may happen on the skill-based, rule-based, or knowledge-based level (Rasmussen 1983). Errors can happen on any of these levels, but it can be hard to distinguish them based on the overt behavior alone. Understanding how users manage their goals and how they come into action is key to understanding user error.

In the following chapter, the focus shifts from the Embodied cognition part of the ETA-triad to the Artifact part.

Chapter 3
Model-Based UI Development (MBUID)

In this chapter[1]:

- The MBUID process allows efficient development of multi-target applications
- As a side-effect, this process produces rich meta-information about the resulting UI (e.g., the intended use patterns as task hierarchies)
- Model-based runtime frameworks allow access to this meta-information through computational introspection

With the rise of mobile appliances and the general move towards ubiquitous computing, user interfaces are no longer bound to specific devices. With new devices and device types constantly appearing, UIs must be adapted (even at run time) to form factors that may have been unforeseen at development time. From the perspective of usability, this poses a big problem. If the UI will be adapted to an unforeseen form factor in the future, then some principles of classical user-centered design (e.g., empirical measurement, see Sect. 1.1) can not be applied.

The ability of user interfaces to adapt to different contexts of use (e.g., new devices) and at the same time being able to preserve their usability has been coined *plasticity* (Coutaz and Calvary 2012). In order to achieve plasticity, Calvary et al. (2003) have proposed a rigorous software engineering process, the so-called CAMELEON reference framework. CAMELEON applies the recommendations of Model Driven Architecture (e.g., the ability to "zoom" in and out between models of different level of abstraction; Miller and Mukerji 2001) to the development of user interfaces. The general idea is to capture the shared properties and functionality of differently adapted UIs in abstract models of these interfaces. The development starts at the highest level

---

[1]Parts of this chapter have already been published in Halbrügge et al. (2016).

© Springer International Publishing AG 2018
M. Halbrügge, *Predicting User Performance and Errors*, T-Labs Series
in Telecommunication Services, DOI 10.1007/978-3-319-60369-8_3

of abstraction. Examples of implementations of the CAMELEON framework are UsiXML (Limbourg et al. 2005) and TERESA (Mori et al. 2004).

3.1   A Development Process for Multi-target Applications

Model-Based UI Development (MBUID; Meixner et al. 2011) specifies informa-tion about the UI and interaction logic within several models that are defined by the designer (Vanderdonckt 2005). The model types that are part of the CAMELEON framework belong to different levels of abstraction. The process starts with a highly abstract task model, e.g., using ConcurTaskTree notation (CTT; Paternò 2003). In contrast to other task analysis techniques, the CTT models contain both user tasks (e.g., data input) and system tasks (e.g., database query). On the next level, an Abstract User Interface (AUI) model is created that specifies platform-independent and modality-independent interactors (e.g., 'choice', 'command', 'output'). At this level, it is still open whether a 'command' interactor will be implemented as a button in a graphical UI or as a voice command. In the following Concrete User Interface (CUI) model, the platform and modality to be used is specified, e.g., a mock-up of a graphical UI. On the last level, the Final User Interface (FUI) is the UI that users actually interact with, e.g., a web page with text input fields for data input and buttons for triggering system actions. The four levels are visualized in Fig. 3.1.

In its original form, the MBUID process targets development time, only. Once the process is completed, no references from the FUI back to the underlying devel-opment models remain. An extension to this approach are runtime architectures for model-based applications (e.g., Clerckx et al. 2004; Sanchez et al. 2008; Blumendorf et al. 2010). These runtime architectures keep the development time models (CTT, AUI, CUI) in the final product and derive the FUI from current information in the underlying models. This allows to adapt the FUI to changes in the models and/or the context of use, thereby reducing complexity during development even further.

Fig. 3.1 Hierarchy of models in the (simplified) CAMELEON reference framework (Calvary et al. 2003)

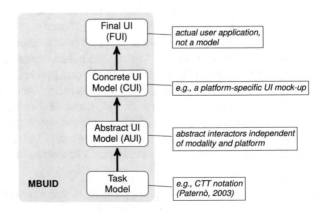

In the context of this work, the most important feature of runtime architectures is the *introspectability* of the FUI. As the underlying models are available at runtime, meta-information about FUI elements like their position in a task sequence (based on the CTT) or their semantic grouping (based on the AUI) can be accessed computationally. Whether and how this meta-information can be exploited for usability predictions will be explored in the main part of this work. The corresponding analysis will be based on a specific runtime architecture that is described in the following section.

3.2 A Runtime Framework for Model-Based Applications: The Multi-access Service Platform and the Kitchen Assistant

A feature-rich example of CAMELEON-conformant runtime architectures is the Multi-Access Service Platform (MASP, Blumendorf et al. 2010). It has been created for the development of multimodal[2] applications in ambient environments like interconnected smart home systems. Within the MASP, a *task model* of the application is available at runtime in ConcurTaskTree format (CTT; Paternò 2003). In addition to the CAMELEON-based *AUI* and *CUI* models, a *domain* model holds the content of an application, i.e., the objects that the elements of the task model can act upon. Information about the current context of use is formalized in a *context* model that is used to derive an adapted final UI at runtime (Blumendorf et al. 2008).

With respect to the overall goal of this work, the MASP architecture has the benefit the derived final UI is mutually linked to its underlying CUI, AUI and task models. How does a MASP-based application actually look like?

The Kitchen Assistant
One reference application of the MASP is a kitchen assistance system (Blumendorf et al. 2008). The kitchen assistant helps preparing a meal for a given number of persons with its searchable recipe library,[3] adapted shopping list generator, and by providing interactive cooking or baking instructions. This application will be used for the empirical analysis of the suitability of MBUID meta-information for automated usability evaluation and the user model development in the following Pt. II of this work. In terms of the ETA-triad (see Chap. 2), the kitchen assistant serves as the Artifact part. A screenshot of its recipe search screen FUI alongside the underlying CTT task model is shown in Fig. 3.2.

---

[2]Graphical UI based on HTML (Raggett et al. 1999) and voice UI based on VoiceXML (McGlashan et al. 2004).

[3]For reference: The recipe library is an example of what the MASP stores in the domain model mentioned earlier.

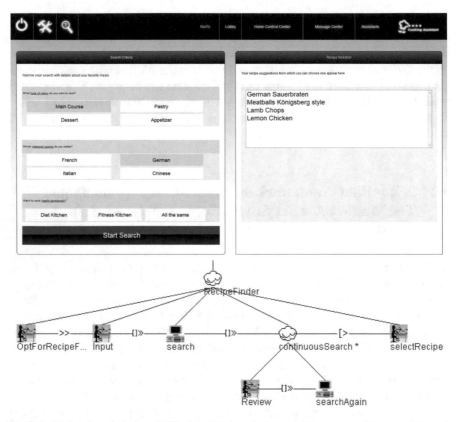

Fig. 3.2 Recipe Search Screen (FUI) of the Kitchen Assistant with corresponding part of the task model (CTT notation, screenshot taken from CTTE, Mori et al., 2002) below

3.3   Conclusion

The current chapter introduced the technical system that will be used throughout this work (i.e., the A part of the ETA triad), a multi-device kitchen assistant based on the Multi-Access Service Platform (MASP; Blumendorf et al. 2010). Being a model-based runtime framework, the MASP provides meta-information about the UI elements of the kitchen assistant through computational introspection. How this property relates to automated usability evaluation is discussed in the next chapter.

Chapter 4
Automated Usability Evaluation (AUE)

In this chapter[1]:

- Maintaining usability is important but costly if empirical evaluation is involved.
- The more devices need to be covered, the costlier the usability evaluation.
- Automated tools based on the psychological characteristics of the users may ease this situation.
- Example: Automated evaluation based on MASP and MeMo (Quade 2015).

The recent explosion of mobile device types and the general move to ubiquitous systems have created the need to develop applications that are equally usable on a wide range of devices. While the MBUID process presented in the previous chapter can ease the development of such applications, the question of the actual usability of these on different devices is still open. Empirical user testing would yield valid answers to this question, but does not scale well if many device targets are to be addressed because time and costs increase (at least) linearly with the number of devices. Automated Usability Evaluation (AUE) may be the proper solution to this problem. In principle, automated tools can be applied to many variations of a UI without additional costs in time. The validity of AUE results is limited, though. In a review, Ivory and Hearst conclude the following:

> It is important to keep in mind that automation of usability evaluation does not capture important qualitative and subjective information (such as user preferences and misconceptions) that can only be unveiled via usability testing, heuristic evaluation, and other standard inquiry methods. Nevertheless, simulation and analytical modeling should be useful for helping designers choose among design alternatives before committing to expensive development costs. (Ivory and Hearst 2001, p. 506)

---

[1] Parts of Sect. 4.1 have already been published in Halbrügge (2016a). Parts of Sect. 4.4 have already been published in Halbrügge et al. (2016).

© Springer International Publishing AG 2018
M. Halbrügge, *Predicting User Performance and Errors*, T-Labs Series
in Telecommunication Services, DOI 10.1007/978-3-319-60369-8_4

Because AUE methods address human behavior towards technological artifacts, their validity depends on how well they capture the specifics of the human sensory-cognitive-motor system (i.e., their psychological soundness).

In the following, the current state-of-the-art in AUE is presented. Afterwards, specific methods for MBUID systems are discussed.

4.1   Theoretical Background: The Model-Human Processor

The application of psychological theory to the domain of HCI has been spearheaded by Card et al.'s seminal book "The Psychology of Human-Computer Interaction". Therein, mainly expert behavior is covered, i.e., when users know how to operate a system and have already formed procedures for the tasks under assessment. Card et al. (1983) are using a computer metaphor to describe such behavior, the so-called model-human processor (MHP). By assigning computation speeds (see cycle times in Fig. 4.1) to three perception, cognition, and motor processors that work together, the MHP is capable of explaining many aspects of human experience and behavior (e.g., lower time bounds for deliberate action, minimum frame rate for video to be perceived as animated vs. a sequence of stills).

Of the step-ladder model introduced in Sect. 2.1, the MHP only covers the lower half. The upper part of the ladder, i.e., knowledge-based behavior, is not addressed by the MHP or the GOMS (Goals, Operators, Methods, and Selection rules) and KLM (Keystroke-Level Model) techniques that are derived from it.

4.1.1   Goals, Operators, Methods, and Selection Rules (GOMS)

The GOMS technique originally aims at explaining how users solve tasks with an artifact and how much time they need to do so. Later extensions add other usability

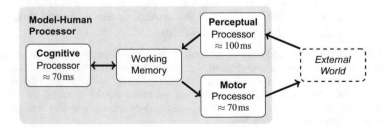

Fig. 4.1 Simplified structure of the Model-Human Processor (MHP; Card et al. 1983). Millisecond values are the average cycle times of the respective processors. Later research assumes 50 ms cycle time for the cognitive processor (John 1990)

aspects like learnability (Kieras 1999). GOMS belongs to the family of task analysis techniques (Kirwan and Ainsworth 1992). User tasks (or *goals*, the G in GOMS) are decomposed into subgoals until a level of detail is reached that corresponds to the three processors in Fig. 4.1. The subgoals on this highest level of detail which are not decomposed anymore are called *operators* (the O in GOMS). Following this rationale, the simple goal of determining the color of a word on the screen yields the following operator sequence:

1. attend-word (cognitive, 50 ms)
2. initiate-eye-movement (cognitive, 50 ms)
3. eye-movement (motor, 180 ms)
4. perceive-color (perceptual, 100 ms)
5. verify-color (cognitive, 50 ms)

The time estimates are given as reported by John (1990). The expected task completion time is the sum of all operator times (430 ms in this case).

In case of (sets of) more complex goals, reusable sequences of operators may emerge which are formalized as *methods* (the M in GOMS). Examples for these are generic methods of cursor movement that are applied during the pursuit of different goals during document editing (e.g., moving a word to another position, deleting a word, fixing a typo in a previous paragraph). If several methods could be applied in the same situation, *selection rules* (the S in GOMS) have to be specified that determine which method to choose.

While GOMS provides fine grained predictions of TCT, it is seldomly applied because it is rather hard to learn (John and Jastrzembski 2010) and corresponding tools are still immature (e.g., Vera et al. 2005; Patton et al. 2012).

4.1.2   The Keystroke-Level Model (KLM)

An easier solution is provided by a simplified version of GOMS, the Keystroke-Level Model (KLM). The KLM mainly predicts task completion times by dividing the necessary work into physical and mental actions. The physical time (e.g., mouse clicks) is predicted based on results from the psychological literature and the mental time is modeled using a generic "Think"-operator M that represents each decision point within an action sequence. Within the KLM, one M takes about 1.35 s, which has been determined empirically by Card et al.. While the generic M operator may oversimplify human cognition, predictions based on the KLM are relatively easy to obtain and are also sufficiently accurate.

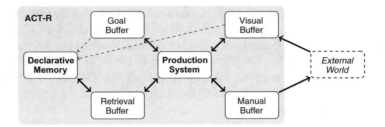

Fig. 4.2 Simplified structure of ACT-R (Anderson and Lebiere 1998; Anderson et al. 2004; Anderson 2007)

4.2   Theoretical Background: ACT-R

A rather theory-driven approach to ensuring cognitive plausibility of an AUE tool is to base it on a cognitive architecture (Gray 2008). Cognitive architectures are software frameworks that incorporate assumptions of the invariant structure of the human mind (e.g., Langley 2016).

A longstanding architecture is the Lisp-based framework ACT-R (Anderson and Lebiere 1998; Anderson et al. 2004; Anderson 2007). The main assumption of ACT-R as a theory is that human knowledge can be divided into declarative and procedural knowledge which are held in distinct memory systems. While declarative knowledge consists of facts about the world (e.g., "birds are animals") which are accessible to conscious reflection and can be easily verbalized, procedural knowledge is used to process declarative knowledge and external inputs from the senses, to make judgements, and to attain goals (e.g., determining whether a previously unseen animal is a bird or not). Contrarily to declarative knowledge, procedural knowledge is hard to verbalize.

In ACT-R, declarative knowledge is modeled using *chunks*, i.e., pieces of knowledge small enough to be processed as a single entity.[2] Procedural knowledge is represented in ACT-R as set of *production rules* which map a set of preconditions to a set of actions to be taken if these conditions are matched.

Taken together, chunks and productions yield a complete Turing machine (Schultheis 2009), which is psychologically implausible. For this reason, ACT-R implements a number of constraints that limit its capabilities. Most importantly, the production rules do not operate directly on declarative memory and sensual inputs, but have communicate with these through small channels ("buffers" in ACT-R nomenclature) that can only hold one chunk at a time. The structure of ACT-R is shown in Fig. 4.2.

In contrast to the MHP, ACT-R as a theory aims at describing and explaining the complete range of behavior and control stages of the step-ladder model described

---

[2]Example: the letter sequence "ABC" can be held in memory as a single chunk by most people who use the Latin alphabet. The arbitrary sequence "TGQ" on the other hand is rather represented as list of three chunks containing individual letters.

in Sect. 2.1. Existing cognitive models using ACT-R on the other hand are often only related to small parts of the complete ladder, e.g., car-driving (mainly skill-based; Salvucci 2006), videocassette recorder programming (mainly rule-based; Gray 2000), or problem solving in math (mainly knowledge-based; Anderson 2005).

4.3 Tools for Predicting Interactive Behavior

How can the theories given above help to predict the usability of software systems? Several tools for (semi-) automated usability evaluation are presented in the following. The presentation is guided by how they cover the three parts of the ETA triad (see Chap. 2); their comparison is done on their scope and on their applicability.

4.3.1 CogTool and CogTool Explorer

Modeling with CogTool (John et al. 2004) aims at predicting task completion times (TCT) for expert users in error free conditions. It is based on the Keystroke-Level Model (Card et al. 1983, see Sect. 4.1) and ACT-R (Anderson et al. 2004, see Sect. 4.2). How the ETA-triad is represented in CogTool is given in Table 4.1.

An important extension has been developed with CogTool Explorer (Teo and John 2008). It implements Pirolli's information seeking theory (Pirolli 1997) to predict exploratory behavior. This allows to model exploratory behavior while interacting with web-based content.

The approach taken by CogTool has proven very successful overall with many applications in different domains (e.g. Distract-R; Salvucci 2009).

Table 4.1 The ETA-triad in CogTool

Embodied cognition
Constraints from embodied cognition provided by ACT-R and the application of the KLM
Task
Tasks are specified by the analyst by performing the corresponding action sequence on a mock-up
Artifact
A mock-up of the Artifact under evaluation has to be created by the analyst. This is a tedious task

Table 4.2 The ETA-triad in GLEAN

Embodied Cognition
Constraints from embodied cognition are implicitly contained in the GOMS model that has to be provided by the analyst and in the MHP-based architecture of GLEAN
Task
Tasks have to be defined by the analyst in a dedicated format
Artifact
The Artifact has to be specified by the analyst in a dedicated format ('Device Specification' in Fig. 4.3)

4.3.2   GOMS Language Evaluation and Analysis (GLEAN)

Another computer tool that is derived from the MHP is GLEAN (GOMS Language Evaluation and ANalysis; Kieras et al. 1995). GLEAN uses a GOMS dialect that specifies user behavior not as sequence of low-level operators (see example in Sect. 4.1.1), but uses a structured language instead that is both easily human-readable and machine-executable. The cognitive architecture used by GLEAN is derived from the MHP, but adds a working memory to the cognitive processor that the models can write to and read from (Table 4.2).

In order to use GLEAN, the analyst must perform a task analysis first (Kieras 1999). A resulting GOMS model together with a dedicated system mock-up and set of tasks can then be used to simulate user behavior with GLEAN. This simulation yields predictions of task completion time, consistency of the UI (operationalized by the re-usability of *methods* within the GOMS model), and a memory workload profile. The structure of GLEAN is given in Fig. 4.3.

While GLEAN provides usability metrics far beyond task completion time, it still does not yield error predictions. This is due to the use of GOMS which lacks interruptability and recovery from error.

4.3.3   Generic Model of Cognitively Plausible User Behavior (GUM)

Butterworth et al. (2000) argue that cognitive models that aim at explaining human behavior are too specific and more abstract user models are better suited for reasoning about interactive systems (which is the original goal of AUE). They propose a solution to this based on formal methods which they call the Generic Model of Cognitively Plausible User Behavior (GUM).

In their approach, the system under evaluation is modeled as a variable holding the *device state*. This can be changed by *actions* which map one device state to another one. The internal state of the user on the other hand is modeled as *belief state* with

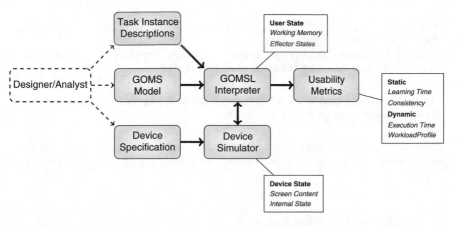

Fig. 4.3 Structure of GLEAN. Adapted from Kieras et al. (1995)

Table 4.3 The ETA-triad in GUM

Embodied Cognition
The GUM is not directly derived from psychological theory, but implements several plausible assumptions, e.g., that users try to advance to the goal and that the users may have incorrect representations ("beliefs") of the system state
Task
The task is specified by its goal state. This is a logical expression that the user wants to evaluate to true. It depends on the belief state of the user (not the actual system state)
Artifact
This is called "device" in this approach. It is modeled as a variable (or set of variables) with system actions that change its state

operations that change the belief. Operations and actions are paired together, i.e., when an action is chosen and the device state is changed accordingly, the corresponding operation changes the user belief as well. Users are assumed to act rationally, i.e., they choose actions that—according to the current belief state—advance the state towards the intended goal (Table 4.3).

Taken together, this is already sufficient to explain postcompletion errors (see Sect. 2.1.2), because no new action will be chosen by the user if the goal expression already evaluates to true. Bolton et al. (2012) have extended the system by adding the possibility of erroneous behavior based on the phenotypical error classification of Hollnagel (1993). The resulting system has been implemented in a formal verification system by Rukšėnas et al. (2014) in order to evaluate the interface design of an infusion pump with good success.

Table 4.4 The ETA-triad in MeMo

Embodied Cognition	

Embodied Cognition

Some aspects of human information processing are phenotypically incorporated into the user interaction model of MeMo, e.g., effects of aging (Schulz 2016)

Task

The task has to be specified by the analyst as user task knowledge, i.e., a set of information items that the user wants to transmit to the system

Artifact

A dedicated mock-up has to be created using the MeMo workbench. Transitions between states are depending on a set of system variables that are dynamically altered during the simulation

4.3.4   The MeMo Workbench

The MeMo workbench (Möller et al. 2006; Jameson et al. 2007; Engelbrecht et al. 2008) differs from the previous approaches in that it targets not only classical GUIs, but also spoken dialog systems (Engelbrecht 2013) and multimodal systems (Schaffer et al. 2015). This is achieved by representing the system under evaluation on a high level of abstraction (Table 4.4). While, e.g., CogTool only uses a storyboard consisting of a sketched mock-up of an application, MeMo needs an additional representation of the application logic using dedicated system variables. Transitions between individual system states in MeMo depend on the current values of these variables (called *conditions* in MeMo) and in turn change these values (called *consequences* in MeMo). MeMo's user simulation progresses towards a given goal state by utilizing a combination of breadth-first and depth-first search on a graph that is created based on the system states and the conditions and consequences of the transitions between them. Additional assumptions about the human users (e.g., less-than-perfect rationality) are incorporated into MeMo's *user interaction model* (e.g., by adding some noise to the current decision).

MeMo has been successfully applied to predict usability issues of spoken dialog systems (e.g., Engelbrecht 2013; Schmidt et al. 2010). More detailed descriptions of MeMo can be found in Engelbrecht (2013, Chap. 2) and Quade (2015, Sect. 4.1.4).

4.4   Using UI Development Models for Automated Evaluation

One major downside of the methods described above is they require a representation of the application in the required format of the respective tool. The development of such a representation is a tedious and time consuming task. As Kieras and Santoro (2004) have put it: "Programming the simulated device is the actual practical bottleneck."

In the case of model-based applications, a promising solution to this "bottleneck" is to exploit the information contained within those models. While the main focus of MBUID is on how adaptable UIs can be *developed* efficiently, the MBUID approach can also help to *evaluate* the usability of the final user interfaces (Sottet et al. 2008).

4.4.1   Inspecting the MBUID Task Model

Paternò and Santoro have proposed an inspection-based evaluation of safety-critical systems based on the task model of such a system which also forms the central part of the MBUID process (Paternò and Santoro 2002). This early approach still lacks both the automation and the predictive value of a full-fledged AUE tool.

In a follow-up paper, Mori and colleagues presented the TERESA tool that uses the task model to deduce problematic states like dead ends or unreachable states within a UI (Mori et al. 2004). As this approach is not grounded in psychological theory, it fails to incorporate the human aspect of error and can not account for typical user errors as discussed in the literature (e.g., Byrne and Davis 2006; Hiltz et al. 2010; Gray 2000, see Chap. 2). TERESA rather targets errors during the formation of the task model as part of the requirements engineering process than errors actual users would make after the release of the product (Table 4.5).

4.4.2   Using Task Models for Error Prediction

Palanque and Basnyat (2004) build upon the task model of an application to predict the tolerance of a system towards user errors. But similar to Paternò and Santoro (2002), their approach consists only of a systematic inspection of the task model while adding a multitude of possible types of errors based on the work of Reason (1990), Rasmussen (1983), and Hollnagel (1998). The inclusion of error theories represents an improvement over Paternò and Santoro's approach, but it still lacks the automatibility that is necessary for an AUE tool (Table 4.6)

Table 4.5   The ETA-triad in TERESA

Embodied Cognition
Not represented
Task
Tasks have to be specified by the analyst
Artifact
The task model (but not the actual UI) of the MBUID process is used. No mock-up is required

Table 4.6 The ETA-triad in Palanque and Basnyat's approach

Embodied cognition
No cognitive theory is used, but phenotypical error types are incorporated
Task
Tasks have to be specified by the analyst
Artifact
The task model (but not the actual UI) of the MBUID process is used. No mock-up is required

4.4.3   Integrating MASP and MeMo

A leading step towards fast and easy AUE for multi-target applications has been developed by Quade (2015). In his work, the MASP runtime framework (Blumendorf et al. 2010, see Chap. 3) is integrated with MeMo (Engelbrecht et al. 2008, see Sect. 4.3.4) and CogTool (John et al. 2004, see Sect. 4.3.1). The resulting structure of the integrated system is shown in Fig. 4.4.

Table 4.7 The ETA-triad in Quade's MASP-MeMo-CogTool system

Embodied cognition
Contraints on human information processing are represented through the incorporation of CogTool (see Sect. 4.3.1)
Task
Have to be specified by the analyst using MeMo syntax (see Sect. 4.3.4)
Artifact
The AUI and the FUI of the model-based application is used. No mock-up is required

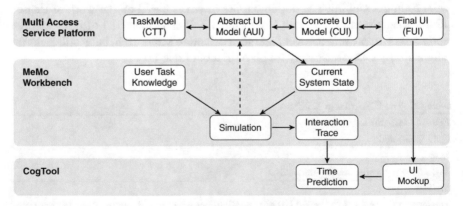

Fig. 4.4 Structure of the integrated MASP-MeMo-CogTool system (Quade 2015). Information flow is denoted by *solid arrows*. The MeMo simulation engine applies interactions on the AUI level of the MASP (*dashed arrow*)

Based on information on both the AUI and FUI level, an internal representation of the current state of the application is provided to MeMo (and CogTool in a later processing step). Thereby, no additional mock-up needs to be created (Table 4.7).

4.5 Conclusion

Maintaining usability is hard, especially for multi-target applications. AUE is a possible solution to this, especially during early development cycles when empirical user testing is not yet possible (Ivory and Hearst 2001). With CogTool (John et al. 2004), GLEAN (Kieras et al. 1995), GUM (Butterworth et al. 2000), and MeMo (Möller et al. 2006), four different generic AUE tools were introduced that all have their individual strengths and weaknesses. Common to all of them is that their application is bound to the creation of a mock-up of the system under evaluation beforehand. In case of applications that adapt or are adapted to many different device targets (i.e., multi-target applications; see Chap. 3), this poses significant overhead.

A solution to this for CAMELEON-conformant multi-target applications is to revert to the shared task model of their different target-specific variants. This approach has been taken in TERESA (Paternò and Santoro 2002) and by Palanque and Basnyat (2004), but their approaches lack the application range and cognitive plausibility provided by the generic tools reviewed before.

A promising candidate for bridging this gap is the MASP-MeMo-CogTool system developed by Quade (2015). Here, a CAMELEON-conformant *runtime* framework provides not only information which is shared across the target-adapted variants (e.g., the task model), but also target-specific information (e.g., UI element placement) to a user simulation engine. This does not only render the creation of UI mock-ups obsolete, it also provides access to the MBUID meta-information that the main research question of this work refers to (see Sect. 1.4). For this reason, Quade's system is used as the technical basis during the analysis of which aspects of this meta-information are suitable for usability predictions that will unfold in the following chapters.

Based on the technical and psychological background presented in the previous three chapters, the research questions of this work can be stated more clearly:

- How does UI meta-information as provided by the MASP relate to human performance and human error? Can such relations be captured with a cognitive user model?
- Can such a cognitive user model provide sufficiently accurate predictions of human interactive behavior with different applications and/or in different contexts?
- How relevant are these predictions for the iterative design of applications? Which aspects of usability can be covered?

Part II
Empirical Results and Model Development

Chapter 5
Introspection-Based Predictions of Human Performance

In this chapter[1]:

- Prediction of the *efficiency* of an interface. How long does a task take?
- UI meta-information can be exploited for improved predictions.
- Access to the MBUID models allows to automate the evaluation process.

Efficiency is an important aspect of the usability of a UI. It can be operationalized as *task completion time* (TCT), i.e., the average time needed to complete a previously specified task, which allows objective measurement. Subjective measures of usability are only weakly connected to task completion time ($R^2 = 0.038$ across 45 studies in a meta-analysis by Hornbæk and Law 2007), which limits the ecological validity of TCT as a predictor of the overall acceptability of an application or a service. Nevertheless, efficiency and TCT play an important role in areas where tasks are mainly routine and time means cost, e.g., business software.

In the context of this work, TCT provides an ideal testbed for checking the main research question, i.e., whether UI meta-information can be used to improve usability predictions, namely for the following reasons:

- TCT can be measured easily (esp. compared to user errors)
- TCT can be explained based on a well-established, simple theory (MHP; Card et al. 1983, see Sect. 4.1)
- Tools exist that allow sufficiently accurate predictions of TCT (CogTool, John et al. 2004, see Sect. 4.3.1). This provides a good baseline for the evaluation of MBUID-based predictions.

This chapter is organized as follows. First, a short theoretical overview of TCT is given. Data from a pretest is used to derive a set of rules for TCT predictions based on information from MBUID models. These rules are then validated on new data.

---

[1] Parts of this chapter have already been published in Quade et al. 2014 and Halbrügge 2016a.

© Springer International Publishing AG 2018
M. Halbrügge, *Predicting User Performance and Errors*, T-Labs Series
in Telecommunication Services, DOI 10.1007/978-3-319-60369-8_5

5.1 Theoretical Background: Display-Based Difference-Reduction

As shown in Chap. 4, the Model-Human Processor and the derived GOMS and KLM techniques allow good predictions of TCTs. But they have also limitations. The main paradigm studied by Card et al. (1983) was document editing and professional secretary work using line-based text editors. Such editors rely heavily on modes (e.g., users cannot add text outside some 'insert' mode) and are operated based on memorized command strings (examples for Vim[2]: `':q!'` or `'10dd'`). Users that behave according to the KLM are assumed to have perfect knowledge in-the-head (Norman 1988). If the visual presentation of the interface does not keep up with their actions, they do not wait for the UI, but type-ahead the next command.

Since then, the software landscape has changed dramatically. Direct manipulation (Frohlich 1997) is the prominent paradigm since the introduction of graphical UIs (GUI), and text processing is now done using WYSIWYG[3] editors. One major effect of these changes is that today's systems aim at providing as much knowledge in-the-world (Norman 1988) as possible. Thereby, the cognitive demand of a task can be significantly lowered. The actual work is shifted from the cognitive to the perceptual domain. As Gray (2000) has pointed out, the presentation of the current state of a system in its GUI reduces the necessity of place-keeping in the users' memory. He has called the resulting user strategy *display-based difference-reduction* (DBDR). When using DBDR, users do not follow a memorized action sequence, but choose their actions based on whether and where the GUI differs (visually) from an intended target state.

According to Gray, this is a satisficing (i.e., non-optimal but good enough) strategy. The application of satisficing strategies like 'hill climbing' (local optimization to attain a global goal, Simon and Newell 1971) or 'difference reduction' (Newell and Simon 1972) is a central aspect of human problem solving. The DBDR strategy is also a vivid example of embodied cognition in the sense that humans use the information provided on a GUI to minimize cognitive load (Wilson 2002).

5.2 Statistical Primer: Goodness-of-Fit Measures

A major prerequisite of an automated usability evaluation system is that its predictions are correct. How can the degree of correctness (or 'goodness of fit') of a predictive model be determined?

Model fits are generally examined by comparing the model predictions to empirical data. Visual examples of such comparisons are given in Figs. 5.1 and 5.2. The most basic fit measure is the root mean squared error (RMSE) which represents an

---

[2]http://www.vim.org/.

[3]What You See Is What You Get.

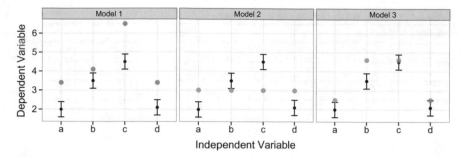

Fig. 5.1 Fit of three hypothetical models to dataset 1. Error bars are 95% confidence intervals

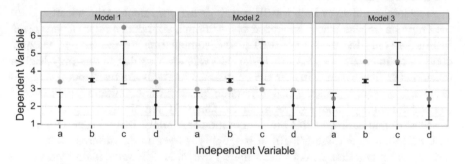

Fig. 5.2 Fit of three hypothetical models to dataset 2. The means are identical to dataset 1. Error bars are 95% CIs. The additional information contained in the confidence intervals is only captured by the MLSD measure

averaged weighted distance between data y_i and model prediction $\hat{y}_i$.

$$\text{RMSE} = \sqrt{\frac{\sum_{i=1}^{n}(\hat{y}_i - y_i)^2}{n}} \qquad (5.1)$$

The RMSE is measured on the same scale as the y_i, and with the same unit. It can thereby be interpreted easily as prediction error of the model, but cannot be compared easily between different scales. It is furthermore highly susceptible to outliers and not very indicative of whether the model captures the variance in the data. An example of this downside is visualized in Fig. 5.1. While Model 1 captures the differences between the four conditions a–d well, the nearly constant Model 2 yields a better RMSE (see Table 5.1).

This is addressed by a complementary goodness-of-fit measure which disregards the scales, namely the squared correlation between data and prediction R^2. This can be interpreted as the proportion of variance in the y_i that is explained by the model. In the hypothetical example, the nearly constant Model 2 explains next to no variance, while Model 1 does so to a great extent. Model 3 finally yields both the

Table 5.1 Goodness of fit of three hypothetical models

Measure	Dataset 1			Dataset 2		
	Model 1	Model 2	Model 3	Model 1	Model 2	Model 3
R^2	0.861	0.001	0.883	0.861	0.001	0.883
RMSE	1.42	1.04	0.64	1.42	1.04	0.64
MLSD	3.0	2.4	1.9	3.5	3.0	6.0

best RMSE and R^2 although it does not discriminate between the conditions b and c of the independent variable.

A general problem of both RMSE and R^2 is that they neglect the uncertainty of the empirical means that they depend on. Although confidence intervals are given in Fig. 5.1, only the means are used in the computation. In the case of unequal confidence intervals (Fig. 5.2), or if the intervals are very large, this may lead to meaningless fit measures. In case the confidence intervals are overlapping to great extents, i.e., there is no statistical difference between the conditions, this can lead to overfitting (cf., Goodfellow et al. 2015, Chap. 5). An alternative goodness-of-fit measure that addresses this problem is the Maximum Likely Scaled Difference (MLSD; Stewart and West 2010). It is roughly[4] based on the differences between data y_i and prediction $\hat{y}_i$, but unlike the RMSE, this difference is divided by the length of the 95% confidence intervals of the y_i. The MLSD is then the maximum of the resulting scaled differences. With cu_i being the upper bound and cl_i the lower bound of the confidence intervals:

$$\text{MLSD} = \max_i \frac{\max(\hat{y}_i - cl_i, cu_i - \hat{y}_i)}{cu_i - cl_i} \qquad (5.2)$$

As the 'true' mean is expected to lie anywhere within cu and cl, the MLSD is set to its theoretical minimum of one even if the $\hat{y}$ lies within the confidence interval. An MLSD close to one therefore means that model fit can not be improved further given the uncertainty (i.e., sampling error) of the empirical data. The impact of this rationale is shown in Fig. 5.2 which features a slightly different dataset than Fig. 5.1. While all means are identical and therefore RMSE and R^2 do not differ between the datasets (Table 5.1), Model 3 no longer yields the best MLSD as it is very bad at capturing condition b.

Fair comparisons between models should also take into account the flexibility (Roberts and Pashler 2000; Veksler et al. 2015) and the (computational) complexity of the models (Halbrügge 2007, 2016b). In the context of this work, this will only be done qualitatively as the main focus lies on the *validity* of the models presented here, not on comparing their fit to other models.

---

[4]More specific: It is based on the difference between the predicted value and its opposite boundary of the confidence interval of the empirical mean.

5.3 Pretest (Experiment 0)

In order to examine whether the KLM (see Sect. 4.1) captures how users interact with the MASP-based kitchen assistant introduced in Sect. 3.2, an initial usability study was conducted.

5.3.1 Method

Participants
This usability test focused on the task of finding a recipe and took place in May 2013. Twenty participants (45% female, $M_{age} = 27.8$, $SD_{age} = 9.6$) were recruited from the paid participant pool of TU Berlin.

Materials
The kitchen assistant was presented on a 19″ (48.3 cm) touch screen mounted on a wall cupboard above the kitchen sink or on a 9.7″ (25 cm) tablet computer (see Fig. 5.3). All interactions of the participants with the user interface were recorded by the MASP system that powers the kitchen assistant and additionally recorded on video to be able to identify system errors or misinterpreted user actions (see placement of camera in Fig. 5.3). Written consent was obtained from all participants.

Design
The study featured an exploratory block using a two-factor split-plot design with the repeated factor being the device used and the other factor being the sequence of user tasks. The participants' attitudes towards the kitchen assistant were assessed with several questionnaires. After completion of the exploratory block, a one-factor between-subjects design was used to assess differences in task completion time between the two devices used.

Procedure
During the exploratory block, the participants solved several tasks of different complexity. This also served as training on the kitchen assistant. During the subsequent experimental block, each participant completed the same five simple tasks on one of the two devices, e.g., "Search for German main dishes and select lamb chops".[5] Task instructions were given verbally by the experimenter.

5.3.2 Results

The video recordings and system logs were synchronized using ELAN (Wittenburg et al. 2006), which was also used to annotate user interaction errors, such as wrong and

---

[5]The full instructions are available at Zenodo (doi:10.5281/zenodo.268596).

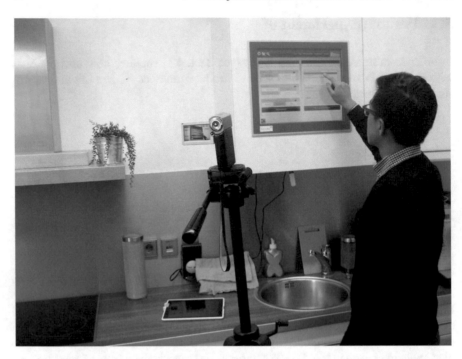

Fig. 5.3 Experimental setup with the kitchen assistant on the mounted touch screen for the pilot study conducted in May 2013. Originally published in Quade et al. (2014). © 2014 Association for Computing Machinery, Inc. Reprinted by permission

unrecognized clicks, as well as system response times, starts, and ends of individual trials.

Because of severe usability problems of the kitchen assistant on the tablet computer, only the trials performed on the mounted screen were used. Data from ten participants (40% female; $M_{age} = 29$, $SD_{age} = 12$) went into the TCT analysis.

Before the actual analysis could start, a classification for the different types of clicks was developed. The class of a click should be related to the time a user needs to perform it. The simplest and fastest one should be repeated clicks on the same UI element. This type will be called *same button* in the following. The other extreme are clicks on buttons that are not part of the same UI screen, i.e., a new page must be loaded before the second click can be performed. This will be denoted as *new screen*. The remaining clicks are performed on the same form, but on different buttons. The buttons of the kitchen assistant's user interface are grouped semantically, e.g., there is a button group called "Regional Dishes" with individual buttons for "French", "German", "Italian", and "Chinese" (see Fig. 5.5). By differentiating between clicks within and across those groups, one finally obtains four types of click pairs, ordered by semantic and also physical proximity: *same button* (repeated clicks); other button in the *same group*; other button in an*other group*; other button on a later displayed *new screen*.

Fig. 5.4 Time per click with original CogTool and extended KLM predictions. Circles denote 20% trimmed means, lines denote bootstrapped 95% confidence intervals (10000 repetitions). Originally published in Quade et al. (2014)

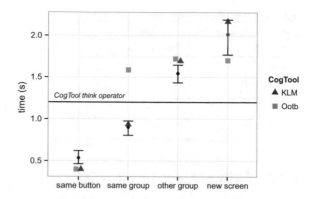

In total 447 single clicks were observed during the experiment. 78 (17%) thereof had to be discarded due to hardware (mainly touch screen) errors. The remaining clicks formed 218 valid pairs of clicks that could be divided into the four categories defined above. The time interval between clicks is significantly different depending on type (linear mixed model with subject as random factor, $F_{3,205} = 19.9$, $p < 0.01$).

Visual examination of the data indicated the presence of extreme outliers (up to 10 seconds between clicks), most probably caused by the inclusion of erroneous trials in the analysis. Any further examination is therefore based on robust statistics like 20% trimmed means (Wilcox 2005).

The comparison to KLM-based predictions was performed by recreating the kitchen assistant and the experimental tasks within CogTool (John et al. 2004). The predicted click times are shown alongside the empirical data in Fig. 5.4.

5.3.3 Discussion

The comparison of the empirical results to the predicted TCT according to the KLM (■ in Fig. 5.6, labeled 'Ootb' for "out-of-the-box") shows interesting deviations. First, the empirical time for the *same group* type of click is substantially shorter than predicted; it even falls below the general 1.2 s M operator time applied by CogTool. Second, the empirical time for the *new screen* type of click is substantially larger than predicted.

5.4 Extended KLM Heuristics

How can the KLM account for the empirical differences? Inspired by the empirical data and backed by the DBDR strategy, three new heuristics for KLM operator placement can be created. The resulting KLM operator sequences for the four click types are given for direct comparison in Table 5.2.

Table 5.2 Effective Operators for the four click types. M = Mental, P = Point, K = Keypress/Touch, L = Look at, SR = System Response time

Click type	Original CogTool	Manual CogTool	Extended KLM
Same button	K	K	K
Same group	M P K	P K	P K L
Other group	M P K	M P K	M P K
New screen	M P K	M P K	SR M P K

5.4.1 Units of Mental Processing

The first observation worth discussing is the difference between the *same group* and *other group* clicks. Buttons in the same semantic group are also physically closer to each other; therefore it should need less time to move the finger from one to another. But is this explanation sufficient? Fitts' Law (Fitts 1954) provides well-researched predictions for the time the user needs for finger pointing movements. Applied to the UI of the kitchen assistant, Fitts' index of difficulty is close to 1 bit within a group and raises up to 3 bit across groups. Even when using a comparatively high coefficient of 150 ms/bit, Fitts' Law only predicts a difference of 300 ms between *same group* and *other group* clicks. This is much less than the 640 ms that was measured; hence Fitts' Law has to be rejected as a single explanation.

In addition, the absolute time between clicks within the same group in the study was approximately 900 ms. If the premise of a generic 1.2 s M operator before every click holds, this would not be possible. It can be concluded that the basic CogTool model does not sufficiently match the data. Therefore, several augmentations of the model based additional rules and heuristics from the literature are presented (John and Jastrzembski 2010).

The original formulation of the KLM forms the basis for the revised model. Card et al. present several rules for the placement of M operators. In principle, those rules are already incorporated in CogTool, but some cannot be applied automatically, i.e., without human *interpretation* of the user interface in question. The rule that is most important for the given experiment is rule 2, which says that if several actions belong to a cognitive unit, only one M is to be placed in front of the first one. This rule definitely applies to the *same button* condition, where one button is clicked a fixed number of times in a row. A more interesting case is the *same group* condition which indicates consecutive clicks within a group of buttons belonging to the same concept, e.g., changing the type of dish from "appetizer" to "dessert". In terms of cognition, this task can be solved using a single chunk of memory that represents the target type of dish ("dessert"), and thus no M operator is to be assigned to the *same group* condition. This also fits well with the empirical mean being noticeably smaller than the 1.2 s M operator time.

5.4.2 System Response Times

Another property of the KLM that can be taken advantage of is the inclusion of system response times. Navigation from one UI screen to another took approximately 500 ms on the hardware setup used in the pilot study, which is comparable to the difference between the *other group* and *new screen* click types. Following the original KLM rules, system response times that occur in parallel to M operators are only taken into account to the extent that they *exceed* the think time, i.e., a frozen system does block mouse and keyboard input, but does not block the mental preparation of the user (Card et al. 1983, p. 263). CogTool applies this rule out of the box, and as the 500 ms screen loading time is shorter than the 1.2 s M operator, CogTool does not predict the difference between the *other group* and the *new screen* conditions (see Fig. 5.6).

The DBDR strategy as introduced above can explain the difference, though. If the users mainly rely on the visual representation of the system state (as opposed to their mental representation of it), then blank time blocks further processing. In the extended model, blank time is therefore no longer applied in parallel to M operator time, but M only starts after the UI is visible to the user.

5.4.3 UI Monitoring

What remains is the 360 ms difference between the *same button* and *same group* click types. Clicking the same button repeatedly does not incorporate movements of the forearm, moving the finger to another button of the same group does. Therefore a difference between the two types is expectable. The movement time can be predicted using Fitts' Law, but again, this does not give sufficiently big estimates. A UI monitoring hypothesis can fill this gap: Given the bad reliability of the touch screen used during the experiment, and again following the rationale of the DBDR strategy, it is highly probable that the participants monitored whether their physical tap on the screen yielded a correct system response (i.e., a visible change of the displayed button). The time that this additional monitoring step needs consists of the time the systems needs to display a change and the time the user needs to notice this change. This system time is about 300 ms for the device used during Experiment 0. CogTool can be used to predict the time the user needs to encode and notice the change.

5.5 MBUID Meta-Information and the Extended KLM Rules

The three heuristics above are of course valuable in themselves, but can they also be used for advancing automated usability evaluation of model-based interfaces? How do they relate to the overall goal of this work?

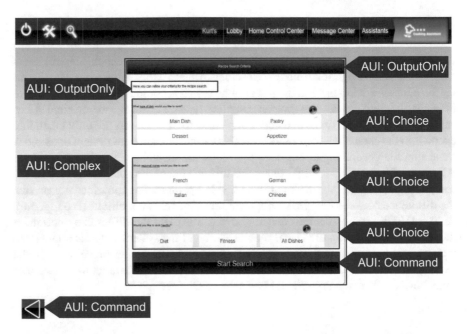

Fig. 5.5 Screenshot of the English version of the kitchen assistant with annotated AUI elements and their types. Originally published in Quade et al. (2014). © 2014 Association for Computing Machinery, Inc. Reprinted by permission

The DBDR-backed second and third heuristic are of general nature and do not need additional information that could be provided by MBUID models. This situation is different for the first heuristic (units of mental processing). The application of this rule in general needs human interpretation of the task and the interface, which would be problematic for the automatibility of the evaluation. Fortunately, the MBUID models are of help. While the search attribute buttons and other buttons on the screen (e.g., for navigating to the home screen) are all belonging to the same class of surface elements, they can be distinguished based on additional information on deeper levels of the MASP system. In the current case, search attribute buttons belong to the AUI type 'choice' (i.e., they are input elements) while navigational buttons are of type 'command' (see Fig. 5.5). And more important: attribute buttons that belong to the same semantic concept (e.g., nationality) are grouped into a single AUI element of type 'complex'.

Based on the AUI types and their hierarchical grouping, the semantic information needed to automate the application of the first heuristic can be extracted from the MBUID system. This is the first proof that MBUID meta-information is actually suitable for the creation of automatable usability predictions.

How the three heuristics can be technically implemented in an actual AUE system is described in Quade (2015).

5.6 Empirical Validation (Experiment 1)

Whether the application of the extended KLM heuristics is actually useful depends on how valid they are. As they have been created *after* the initial data collection in Experiment 0, their validity still needs to be established. This was approached with a dedicated validation study with new participants and additional user tasks.

5.6.1 Method

Participants
The study took place in November 2013, 12 participants were recruited mainly from the group of student workers of Technische Universität Berlin (17% female; $M_{age} = 28.8$, $SD_{age} = 2.4$).

Materials
The presentation of the user interface was moved from the wall-mounted touch screen to a personal computer with integrated 27″ (68.6 cm) touch screen with a 16:10 ratio. This was done both to reduce error rates compared to the first experiment and to test whether the model generalizes well to new devices due to adaptation of the UI caused by a different aspect ratio and size of screen.

Design
A single-factor within-subjects design was applied. The independent variable was the click type on four levels (same button, same group, other group, new screen) as defined in Sect. 5.3. The dependent variable was the time the participants needed to perform clicks of these four types.

Procedure
In order to achieve a higher coverage of the kitchen assistant's functionality, each participant completed 34 individual tasks in five randomly ordered blocks.

Besides the changes in participant group, physical device, and task selection, the experimental procedure of the previous study was closely followed.[6] User actions were again logged by the MASP and recorded on video. In the same way, system response times, start times and end times of the individual tasks were annotated using ELAN (Wittenburg et al. 2006).

5.6.2 Results

A total of 180 min of video footage was recorded, about six times the amount of Experiment 0. For being able to compare the results with the previous experiment,

---

[6]The full instructions are available at Zenodo (doi:10.5281/zenodo.268596).

extended KLM predictions (Experiment 1), originally published in Quade et al. (2014)

an analysis on the click-to-click level was conducted first. A total of 1930 pairs of clicks that can be divided into the classification introduced in Sect. 5.3 were observed. Means and confidence intervals for these are given in Fig. 5.6. The differences found in Experiment 0 are qualitatively replicated in the current study. There was an overall increase in speed, though. The average time between clicks decreased from 1.41 s to 1.04 s between the experiments (linear mixed model (Pinheiro et al. 2013), additional factors: click type as fixed and subject as random effect, $F_{1,20} = 55.5$, $p <)0.01$).

5.6.3 Discussion

The results of the validation experiment are qualitatively similar to the first study (Sect. 5.3), but there are also big quantitative differences. The overall gain in click speed between the experiments could be accounted to differences in user group, device used, length of the experiment and so on. As all of these variables are confounded, they can neither be confirmed nor rejected as influence factors.

What is more important is the degree of generalization of the cognitive model to the changed situation. The goodness-of-fit of the model to the new data is an important indication of the usefulness of the approach to automated usability evaluation in general. This line of thought will be taken up again in the chapter discussion below (Sect. 5.8).

Of the fit statistics applied, RMSE and maximum relative difference are most sensitive to overall shifts in the data, whereas the determination coefficient R^2 neglects these and identifies changed relations between (classes of) observations (see Sect. 5.2). R^2 being very high in the current case means that especially the KLM model describes the differences between the identified types of clicks very well. RMSE, scaled difference (MLSD) and relative difference on the other hand clearly show that the model predictions miss the actual task completion times.

Table 5.3 Time per click compared to KLM predictions (Experiments 0–4)

Experiment	Date	N	Original CogTool			Extended KLM		
			R^2	RMSE	MLSD	R^2	RMSE	MLSD
0 Original pretest	05.2013	10	.597	0.39 s	4.6	.995	0.13 s	1.4
1 Original validation	11.2013	12	.425	0.61 s	20.0	.927	0.47 s	13.8
2 Error modeling	07.2014	20	.485	0.43 s	10.8	.930	0.25 s	10.2
3 Eye-tracking	01.2015	24	.440	0.51 s	16.3	.904	0.32 s	15.7
4 Multi-tasking	05.2015	12	.595	0.43 s	6.0	.975	0.21 s	5.6

One promising result is that the modified KLM model still yields substantially higher fits than CogTool out of the box in the second study (see Table 5.3 on p. 50). This provides backing for the validity of the hypotheses that were derived from the results of the Experiment 0. The goodness-of-fit computed on task completion times being comparable to the one computed on click level also hints at the robustness of the general approach.

Finally the most general question: Does the relatively bad fit put the KLM that formed the theoretical basis of the user model into question? Looking at Fig. 5.6 shows that nearly all clicks moved below that 1.2 s M operator time in the validation study. This could be taken as evidence for the general inappropriateness of such an operator. While being correct in itself, this argumentation misses the heuristic nature of the KLM. As Card et al. pointed out, users differ a lot in how they mentally encode a task, and higher levels of user expertise can be modeled by placing fewer M operators for the same task (Card et al. 1983, p. 265). Taken together with the highly selective group of participants in the second study, this can explain the apparent disappearance of mental preparation times.

5.7 Further Validation (Experiments 2–4)

In the next chapters, three additional experiments will be presented that use the kitchen assistant. These experiments differ from Experiment 1 in several respects, e.g., the devices used (Experiment 2, Sect. 6.5), additional data recorded (Experiment 3, Sect. 6.7), and the presence of additional tasks (Experiment 4, Sect. 6.8). Experiments 2 and 3 feature a newly implemented "simple" version of the UI of the kitchen assistant that targets handheld devices. None of these experiments has beenexplicitly designed for the replication of the TCT analysis above. They should

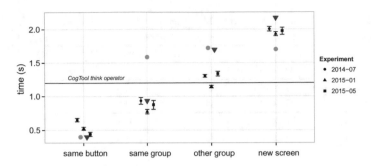

Fig. 5.7 Time per click with CogTool and extended KLM predictions. ● denote original CogTool predictions, ▼ denote extended KLM predictions

therefore provide a good opportunity to test the generalizability of the KLM extensions.

The validity of the extended heuristics is examined by comparing the goodness-of-fit of classical KLM predictions (John et al. 2004, using CogTool;) with the ones obtained after applying the new heuristics (see Table 5.3 and Fig. 5.7). Along with R^2 and RMSE, the Maximum Likely Scaled Difference (MLSD; Stewart and West 2010) is applied as goodness-of-fit measure (see Sect. 5.2). A model that achieves a MLSD close to one should therefore not be refined any further as this would bear the risk of overfitting.

5.8   Discussion

The extended KLM rules proposed in this chapter yielded good empirical fits to new data. The resulting model generalizes well to different experimental paradigms. Compared to the current state-of-the art (CogTool, John et al. 2004), the extended model provides significantly improved prediction accuracies.

At the same time, the automatic export from MeMo provides much higher convenience compared to CogTool (no UI mock-up required; Quade 2015). An overview of how MBUID meta-information is applied by the system is given in Fig. 5.8.

While being well-situated in psychological theory, the generality of the extended rules may be limited as their empirical basis is small and confined to the kitchen assistant. Further validation with larger samples and other applications is needed.

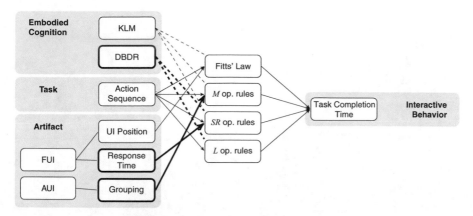

Fig. 5.8 Overview over the UI meta-information used by the extended KLM model together with the processes that link them to interactive behavior. *Thick lines and arrows* Additional information flow in the MASP-MeMo system that allows improved predictions compared to the original CogTool simulation (*thin lines*)

5.9   Conclusion

The current chapter explored how UI meta-information can help to automatically predict the efficiency of a UI with respect to a given set of tasks. Efficiency is operationalized as the time needed to complete those tasks; the theoretical background for the predictions is provided by the KLM (Card et al. 1983). Results from a small pretest suggested that the KLM as formulated in the 1980s does not capture the aspects of current interaction paradigms well. The rise of direct manipulation and WYSIWYG has led to user strategies that rely more on the visual information presented in a UI than on their internal representation of it (namely 'display-based difference-reduction', Gray 2000). This led to the formulation of two new KLM heuristics

- blank time (i.e., no visual information available) blocks the user
- users monitor the visual feedback of their actions

The new heuristics were implemented in CogTool (John et al. 2004). Together with the automatic application of the existing 'cognitive unit' heuristic (if a string of actions belong to *one cognitive unit*, then apply only one *M* operator for the complete string of actions, Card et al. 1983), this lead to substantially better fits. The good fit of the semi-automatic system appears to be robust across a range of experiments that use the kitchen assistant.

The results from the current chapter thereby provide a first answer to the main research question. It is actually possible to exploit UI meta-information to predict certain aspects of the usability (here: efficiency) of the UI. The following chapter will expand this to UI effectiveness, i.e., predictions of user errors.

Chapter 6
Explaining and Predicting Sequential Error in HCI with Cognitive User Models

In this chapter[1]:

- Predict the *effectiveness* of an interface. Is it error-prone?
- Prerequisite: A theoretical account of action control and procedural error
- Implementation as an executable user model in ACT-R (Anderson et al. 2004)
- Together with the user model, UI meta-information can be used to predict error rates for different UI elements.

While the results on introspection-based TCT prediction reported in the previous chapter are encouraging, they suffer from the reduced ecological validity that is inherent to the TCT measure. Not only that TCT is only weakly connected to the attitudes of potential users towards an interface (see discussion in Chap. 5 on page 37). When it comes to comparisons between subsequent iterations of a single UI design, TCT predictions usually differ only in the range of hundreds of milliseconds (cmp. Fig. 5.7).

The current chapter will therefore explore the relationship between UI meta-information and a more significant aspect of usability: human error. While the connection between errors and subjective measures of usability is again very low ($R^2 = 0.027$ across 39 studies in a meta-analysis by Hornbæk and Law 2007), the potential impact of errors is much higher compared to the impact of a few hundred milliseconds time delay. Potential damages range from prolonged task execution to data loss to severe system failure and death (e.g., in the medical domain or in a power

---

[1]Parts of Sects. 6.3 and 6.4 have already been published in Halbrügge and Engelbrecht (2014). Parts of Sects. 6.5 and 6.6 have already been published in Halbrügge et al. (2015b). Parts of Sect. 6.7 have already been published in Halbrügge et al. (2016). Parts of Sect. 6.8 have already been published in Halbrügge and Russwinkel (2016).

© Springer International Publishing AG 2018
M. Halbrügge, *Predicting User Performance and Errors*, T-Labs Series in Telecommunication Services, DOI 10.1007/978-3-319-60369-8_6

plant control room scenario; Reason 2016). Human error is therefore an important, but insufficiently researched factor of usability. In contrast to TCTs which are understood and can be predicted quite well, no commonly accepted theory exists in the domain of errors. Two decades ago, John and Kieras stated:

> No methodology for predicting when and what errors users will make as a function of interface design has yet been developed and recognized as satisfactory ... even the theoretical analysis of human error is still in its infancy. (John and Kieras 1996b, p. 301)

As of today, this situation has not improved much. The variety of error classification schemes presented in Chap. 2 should be sufficient to prove this point.

This chapter will therefore start with a review of a promising theoretical concept that relates interface design to error proneness (goal relevance; Ament 2011b). Whether goal relevance has an impact on user behavior in general will first be tested using empirical data from the previous chapter. The validity of the underlying theory will then be assessed using cognitive modeling with ACT-R (Anderson et al. 2004, see Sect. 4.2).

The following sections shift the focus to human error in particular. Newly gathered data show that the theoretical model from the psychology lab is not sufficient to explain the error patterns that users exhibit when interacting with the kitchen assistant in a real-world scenario. With *task necessity*, a second concept that relates UI design to error proneness is derived from the data and a new cognitive process is proposed that extends the memory-based MFG model with a visual cue-seeking strategy. The validity of these additions is assessed with two subsequent experiments.

The chapter closes with a review of both the empirical findings and theoretical derivations with regard to their instrumentality for automated usability evaluation of model-based user interfaces.

6.1   Theoretical Background: Goal Relevance as Predictor of Procedural Error

The best known examples of procedural error during system use are postcompletion errors (e.g., forgetting the originals in the copy machine; Byrne and Davis 2006, see Sect. 2.1.2) and initialization errors (e.g., forgetting to reset Caps Lock before typing a password; Gray 2000). Common to both of them is that these errors happen during procedural steps that do not directly contribute to the users' actual goals (i.e., making copies; logging into a system). This common property of goal-irrelevance of a subtask has been coined *device-orientation* (Ament 2011a; Gray 2000), its opposite is analogously called *task-orientation*.[2] The concept of device-orientation is based on

---

[2]This is not identical to the concept of task-orientation that is used by some work psychologists (e.g., Ulich et al. 1991).

the MFG by assuming that device-oriented tasks are "more weakly represented in memory" (Ament et al. 2013). In her dissertation, Ament concludes that

> Device-oriented steps do not make a direct contribution towards the task goal but are required for the operation of the device, while task-oriented steps bring the user directly closer to their goal. Device-oriented steps are more problematic than their task-oriented counterparts, because the activation levels on these steps are lower. Therefore, these steps should be designed out of devices as much as possible. (Ament 2011b, p. 166)

The actual term originates from work by Cox and Young (2000); a similar concept is Kirschenbaum's 'task-tool ratio' (Kirschenbaum et al. 1996). There is also an interesting overlap with views from applied domains: Raskin (1997) popularized the complaint that "a dialog box that has no choices (e.g., you can only press ENTER before you can do any other task) has a productivity of 0", because the user cannot transfer any knowledge to the system using it. Raskin's information theoretic concept fits nicely with notion of device-orientation used in this work: such tasks do not convey any information that is specific to the user's current goal.

6.2   Statistical Primer: Odds Ratios (OR)

The effects of different independent variables on various kinds of errors are analyzed in this chapter using generalized linear mixed models (GLMM; Bates et al. 2013). Assessing the strength of association with a binary variable like erroneous versus correct selection is not easy. The approach taken by GLMMs is to convert the relative frequency of the binary outcomes to *odds* (e.g., $p = 0.5$ yields $1 : 1 = 1$; $p = 0.1$ yields $1 : 9 \approx 0.111$) and to assess the influence of an independent variable by computing the ratio of these odds.

The resulting *odds ratios* (OR) can theoretically reach from zero to infinity. An OR of one indicates statistical independence of the two variables. The larger an OR above one (up to infinity), the stronger the positive relationship (larger values of the independent variable lead to higher probability of the binary outcome). The smaller an OR below one (down to zero), the stronger the negative relationship (larger values of the independent variable lead to lower probability of the binary outcome).

It is important to notice that ORs can not be interpreted as relative risk, i.e., an OR of two does not mean that the outcome is twice as likely (Davies et al. 1998). If the base probability of the binary outcome is very low, ORs are numerically close to their corresponding relative risks, though. In the following analyses, ORs are mainly given as estimators of the strength of the influence (effect size) of different variables on error probabilities. See Agresti (2014) for a more detailed description and discussion.

6.3   TCT Effect of Goal Relevance: Reanalysis of Experiment 1

Given the MFG theory (Sect. 2.3.3) and the goal relevance concept introduced above, it shall now be assessed whether these are suited for the overall goal of this work, i.e., whether goal relevance is instrumental for error predictions within an automated usability evaluation. As a first step, already existing data from Experiment 1 (Sect. 5.6) will be analyzed with this new focus. The individual trials of Experiment 1 can be grouped into five phases:

1. Listening to and memorizing the instructions for the given trial.
2. Entering the search criteria (e.g. "German" and "Main dish") by clicking on respective buttons on the screen. This could also contain deselecting criteria from previous trials.
3. Initiating the search using a dedicated "Search" button. This also initiated switching to a new screen containing the search results list if this list was not present, yet.
4. Selecting the target recipe (e.g. "Lamb chops") in the search results list.
5. Answering a simple question about the recipe (e.g. "What is the preparation time?") as displayed by the kitchen assistant after having selected the recipe.

The first and last phase were not analyzed as they do not create observable clicks on the touch screen. Of the remaining three phases, entering search criteria and recipe selection are task-oriented, while the intermediate "Search"-click is device-oriented.

6.3.1   Method

A total of 18 user errors were observed during Experiment 1. Nine were intrusions and nine were omissions. The application logic of the kitchen assistant inhibits overt errors during the device-oriented step. Due to the small number of overt errors, task completion time was chosen as dependent variable and all erroneous trials were excluded from this first analysis.

As the MFG is based on memory effects, the focus of the analysis is laid on steps that task only the memory and motor system. All subtasks that need visual search and encoding were excluded (phase 4: searching for the target recipe in the results list and clicking on it); the same was applied to steps that incorporate substantial computer system response times (i.e., moving to another UI screen).

Table 6.1 Regression coefficients (coef.) with confidence intervals (CI) and analysis of variance results for Experiment 1. Individual slopes for Fitts' Index of Difficulty (ID) ranged from 121–210 ms/bit

Factor name	coef. (ms)	95% CI of coef. (ms)	$F_{1,802}$	p
Fitts' ID	165	126 to 204	111.1	<.001**
Trial block	−55	−71 to −39	45.9	<.001**
Device-oriented step	104	53 to 154	16.4	<.001**

Note **highly significant (p < .01); *significant (p < .05); †marginally significant (p < .10)

6.3.2 Results

817 clicks remained for further analysis; 361 (44%) of these were device-oriented. The average time to perform a click was 764 ms (SD = 381) for task-oriented and 977 ms (SD = 377) for device-oriented steps.

As the kitchen assistant has been created for research in an area different from HCI, it introduces interfering variables that need to be controlled. The motor time needed to perform a click on a target element (i.e., button) depends strongly on the size and distance of the target as formalized in Fitts' law (Fitts 1954). Fitts' index of difficulty (ID) can not be held constant for the different types of clicks, it was therefore introduced into the analysis as covariate. As the click speed (i.e., Fitts' law parameters) differs between participants, linear mixed models (LMM; Pinheiro et al. 2013) with subject as random factor and Fitts' law intercept and slope within subject are applied.

There also was a small, but consistent speed-up during the course of the experiment that led to the introduction of the trial block as additional interfering variable. The analysis of variance was conducted using R (R Core Team 2014). All three factors yielded significant results, there was a prolongation effect for device-oriented steps of 104 ms. The results are summarized in Table 6.1.

6.3.3 Discussion

The main objective of the analysis is met, as a significant execution time delay for device-oriented steps could be identified. This means that the concept of goal relevance is in principle relevant for automated usability evaluations. What remains open is whether the effect can be captured with a computational user model that would be part of such an evaluation system. This is equivalent to the question whether the corresponding cognitive mechanism proposed by the MFG (i.e., lack of activation) can account for the time delay. The next section addresses this question.

6.4   A Cognitive Model of Sequential Action and Goal Relevance

As the concept of goal relevance builds upon the memory for goals theory (Altmann and Trafton 2002, see Sect. 2.3.3), which is in turn derived from ACT-R (the theory, not the computational implementation), the cognitive model was implemented based on the cognitive architecture ACT-R (Anderson et al. 2004). Within ACT-R, memory decay is implemented based on a numerical activation property belonging to every chunk (i.e., piece of knowledge) in declarative memory. Associative priming is added by a mechanism called *spreading activation*.

This led to the translation of the tasks used in the experiment into chains of goal chunks. Every goal chunk represents one step towards the target state of the current task. One element of the goal chunk ("slot" in ACT-R speak) acts as a pointer to the next action to be taken. After completion of the current step, this pointer is used to retrieve the following goal chunk from declarative memory. The time required for this retrieval depends on the activation of the chunk to be retrieved. If the activation is too low, the retrieval may fail completely, resulting in an overt error.

The cognitive model receives the task instructions through the auditory system, just like the human participants did. For reasons of simplicity, the task information was reduced as much as possible. The user instruction "Search for German dishes and select lamb chops" for example translates to the model instruction "German on; search push; lamb-chops on". The model uses this information to create the necessary goal chunks in declarative memory. No structural information about the kitchen assistant is hard coded into the model, only the distinction that some buttons need to be toggled on, while others need to be pushed.

While the model should in principle be able to complete the recipe search tasks of the experiment with the procedural knowledge described above, it actually breaks down due to lack of activation. Using unaltered ACT-R memory parameters, the activation of the goal chunks is too low to be able to reach the target state (i.e., recipe) of a given task. The goals therefore need strengthening and spreading activation is the ACT-R mechanism that can provide it. The application of spreading activation is inspired by close observation of one of the participants who used self-vocalization for memorizing the current task information. The self-vocalization contained only the most relevant parts of the task, which happen to be identical to the task-oriented steps of the procedure. This implies that goal chunks that represent task-oriented steps receive more spreading activation than their device-oriented counterparts. This will be called the *task priming assumption* in the following. This assumption is in line with the discussion of postcompletion errors on the basis of the memory for goals model in Altmann and Trafton (2002) and the general concept of device-orientation with respect to activation in Ament et al. (2010). A schematic flow chart of the model[3] is given in Fig. 6.1.

---

[3]The source code of the model is available on GitHub (doi:10.5281/zenodo.53197).

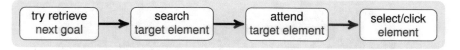

Fig. 6.1   Schematic flow chart of the cognitive model

6.4.1   Model Fit

For the evaluation of the model, ACT-CV (Halbrügge 2013) was used to connect it directly to the HTML-based user interface of the kitchen assistant. In order to be able to study the effect of spreading activation in isolation, activation noise was disabled and only the value of the ACT-R parameter that controls the maximum amount of spreading activation (mas) was manipulated. The higher this parameter, the more additional activation is possible.

The overall fit of the model was evaluated by dividing the clicks into eight groups by the screen areas of the origin and target click position (e.g., from type of dish to search; from search to recipe selection) and comparing the average click times per group between the human sample and the model. Besides the traditional goodness-of-fit measures R^2 and root mean squared error (RMSE), the maximum likely scaled difference (MLSD; Stewart and West 2010, see Sect. 5.2) was applied. The relative difference between the empirical means and the model predictions is given in percent (%diff). The results for five different amounts of activation spreading are given in Table 6.2.

The model is overall slower than the human participants, resulting in moderately high values for RMSE, MLSD, and relative difference. The explained variance (R^2) on the other hand is very promising and hints at the model capturing the differences between different clicks quite well.

Table 6.2   Average click time (M_{time}), average memory retrieval time (M_{mem}), determination coefficient (R^2), root mean squared error (RMSE), maximum likely scaled difference (MLSD), and maximum relative difference (%diff) for different amounts of activation spreading (mas)

mas	M_{time} (ms)	M_{mem} (ms)	R^2	RMSE (ms)	MLSD	%diff (%)
2	1785	591	0.759	982	16.5	66
4	1509	315	0.738	687	12.1	58
6	1291	99	0.881	477	8.5	50
8	1231	37	0.912	422	7.9	48
10	1210	15	0.893	406	7.8	48

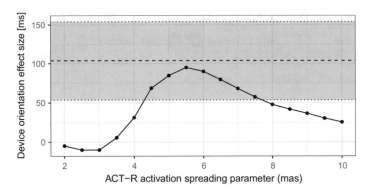

Fig. 6.2 Device orientation effect size depending on spreading activation amount. The *gray* area between the *dotted lines* demarks the 95% confidence interval of the effect in Experiment 1

6.4.2   Sensitivity and Necessity Analysis

In order to test whether the model also displays the goal relevance effect, a statistical analysis identical to the one applied to the human data was performed and the resulting regression coefficients were compared. While an acceptable fit of the model is necessary to support the activation spreading hypothesis, it is not *sufficient* to prove it. By manipulating the amount of activation spreading, a sensitivity and necessity analysis can be performed that provides additional insight about the consequences of the theoretical assumptions (Gluck et al. 2010). Average coefficients from a total of 400 model runs are displayed in Fig. 6.2. It shows an inverted U-shaped relationship between spreading activation and the goal relevance effect. For intermediate spreading activation values, the time delay predicted by the model falls within the confidence interval of the empirical coefficient, meaning perfect fit given the uncertainty in the data.

6.4.3   Discussion

The MFG model is able to replicate the effects that were found in Experiment 1. The model being overall slower than the human participants could be caused by the rather low Fitts' law parameter used within ACT-R (100 ms/bit) compared to the 165 ms/bit that were observed empirically.

Spreading activation is not only necessary for the model to be able to complete the tasks, but also to display the goal relevance effect (Fig. 6.2). The task priming assumption can therefore be considered a sound explanation of the disadvantage

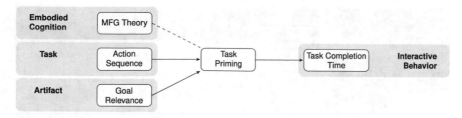

Fig. 6.3   Task completion time as a function of goal relevance

of device-oriented steps. Too much spreading activation reduces the effect again, though. This can be explained by a ceiling effect: The average retrieval time gets close to zero for high values of mas (M_{mem} in Table 6.2), thereby diminishing the possibility for timing differences.

How relevant is a 100 ms difference in real life? Probably not too much by itself. What makes it important is its connection to user errors. Errors itself are hard to provoke in the lab without adding secondary tasks that interrupt the user or create strong working memory strain, thereby substantially lowering external validity.

With regard to the instrumentality of goal relevance in the context of automated usability evaluation, it can be noted that not only there is an effect of goal relevance (as analyzed in the previous Sect. 6.3), this effect is also captured well by a cognitive model based on the MFG theory. This means that goal relevance as a concept can be used in an AUE system, given that the AUE system is capable of identifying device-oriented UI elements. The corresponding information flow is visualized in Fig. 6.3.

The empirical basis of Experiment 1 however is too weak to allow a proper analysis of the relationship between goal relevance and human error. This will be addressed with a new experiment in the following section.

6.5   Errors as a Function of Goal Relevance and Task Necessity (Experiment 2)

After the goal relevance of a user task has been identified as an influencing factor on the usability of a system, several questions remain open:

- the influence of goal relevance on errors
- the connection to UI meta-information
- the generalizability of goal relevance effects across UIs and devices.

Therefore, a new experiment was designed that should address these questions.

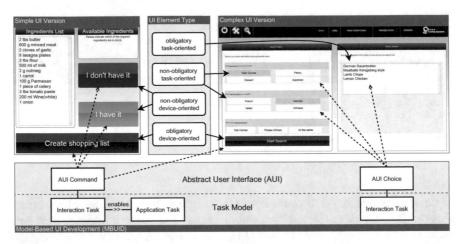

Fig. 6.4 Screenshots of the kitchen assistant. Ingredients list of the simple UI on the *left*, recipe search of the complex UI with two screens side-by-side on the *right*. UI element type is indicated by *solid arrows*, AUI type indicated by *dashed arrows*

The scarcity (yet pervasiveness) of procedural errors during routine tasks (Reason 1990) makes statistical analysis difficult. Researchers have used secondary tasks (e.g., Byrne and Davis 2006; Ruh et al. 2010) or interruptions (e.g., Li et al. 2008; Altmann et al. 2014; Botvinick and Bylsma 2005) to increase error rates. In this work, both options are rejected for reasons of ecological validity, and repeated measures and a medium sized samples are used instead.

Experiment 2 focuses on procedural errors during the usage of the kitchen assistant from Sect. 3.2. In order to explore the multi-target capabilities of the kitchen assistant, a personal computer with large touch screen and a tablet computer were chosen as physical devices for the experiment. To match the characteristics of each device, two versions of the user interface of the kitchen assistant were created. The first version is a tablet-oriented, *simple* one, optimized for portrait mode, with larger buttons and fonts, and rather few elements per screen. The other UI version is more *complex*, using smaller buttons and fonts so that more elements fit on the screen. Some screens of the simple UI are shown side-by-side in the complex version, thereby reducing the necessity of navigation between screens. The complex UI targets the large personal computer and is optimized for landscape mode. Annotated screenshots of the two UI versions are shown in Fig. 6.4.

Alongside goal relevance, a second property of a UI elements was added to the experimental design: *task necessity*, i.e., whether interacting with an element is oblig-atory or not. Making tasks obligatory is a common interaction design strategy to circumvent the downsides of device-orientation. Most teller machines, e.g., hand out the money only if the bank card has been taken by the user before. Thereby, the device-oriented step (taking the bank card) is made obligatory and the risk of users forgetting it is reduced.

Experiment 2 sheds light on how goal relevance and task necessity influence the probability of procedural errors while interacting with the kitchen assistant.

6.5.1   Method

Participants
The experiment was conducted in July and August 2014. Twenty members of the Technische Universität Berlin paid participant pool took part. There were 5 men and 15 women, with an age range from 18 to 59 (M = 32.3, SD = 11.9). As the instructions were given in German, only fluent German speakers were allowed to take part. Written consent was obtained from all participants.

Materials
The experiment was conducted in a fully equipped lab kitchen at Technische Universität Berlin (see Fig. 5.3). A personal computer with 27″ (68.6 cm, landscape mode) touch screen and a 10″ (25.7 cm, portrait mode) tablet, both placed near the sink, were used to display the interface of the kitchen assistant. All user actions were recorded by the computer system. The participants' performance was additionally recorded on videotape for subsequent error identification.

Design
A four-factor within-subjects design was applied, the factors being UI version (simple versus complex), physical device (screen versus tablet), goal relevance (device-vs. task-orientation), and task necessity (non-obligatory vs. obligatory). Every participant went through all four combinations of UI version and physical device in randomized order. User tasks were analogously grouped into four blocks of eleven to twelve individual tasks. Block orders were counterbalanced across participants as well.

The combination of the goal relevance and task necessity factors is called *UI element type* in the following. While goal relevance can be derived from the MBUID models (see Sect. 8.1), task necessity is only implicitly represented in the assistant's interaction logic. Mandatory subtasks are related to all elements that lead to the next screen or unhide buttons on the current screen, the latter being the case for the selection of a recipe in the search result list (see Fig. 3.2).

Procedure
Every block started with comparatively easy recipe search tasks, e.g., "search for German main dishes and select lamb chops".[4] Users would then have to change the

---

[4]English translations of the actual instructions are given here for reasons of comprehensibility. The original instructions are available at Zenodo (doi:10.5281/zenodo.268596).

Table 6.3 Mixed logit model results for Experiment 2. OR: Odds ratio for errors

Factor name	OR	95% CI of OR		z	p
Device (Tablet vs. Screen)	0.74	0.49	1.12	−1.42	.156
UI variant (Simple vs. Complex)	1.17	0.78	1.77	0.75	.452
Device-oriented step	0.29	0.15	0.58	−3.63	<.001**
Non-obligatory step	0.75	0.43	1.34	−1.02	.307
Interaction Dev.-or. × Non-obligatory	6.76	2.57	16.98	3.99	<.001**

Note **highly significant (p < .01); *significant (p < .05); †marginally significant (p < .10)

search attributes, e.g., "change the dish from appetizer to dessert and select baked apples". The second half of each block was made of more complex tasks that were spread over more screens of the interface and/or needed memorizing more items. The participants either had to create ingredients lists for a given number of servings, or had to make shopping lists using a subset of the ingredients list, e.g., without salt and flour. All instructions were read to the participants by the experimenter. Every individual trial was closed by a simple question the participants had to answer to keep them focused on the kitchen setting, e.g., "how long does the preparation take?" With an initial training phase and exit questions the whole procedure took less than an hour.

6.5.2   Results

Errors were identified by comparing the observed click sequences with optimal ones (see definition of procedural error in Sect. 2.1). Whenever a step of the optimal sequence was missing, this was recorded as an *omission*. Unnecessary additional steps performed by the participants were analogously recorded as *intrusions*. A special case of intrusions are *perseverations*, when an action is erroneously repeated.

In total, 6359 clicks were observed. 104 (1.6%) of these were classified as user errors. 56 (0.9%) were omissions, 38 (0.4%) were intrusions, and 10 (0.02%) were perseveration errors. On trial level, 9.3% of the user tasks were not completed correctly by the participants on the first try.

The perseveration errors were bound to a button for increasing the number of servings that needed several consecutive presses. As the video recordings clearly showed that all perseverations were not caused by the users, but by occasional excesses of UI lag, they were removed from further analysis.

Due to the scarcity of errors and the use of repeated measures, χ^2-based significance tests could not be used (Agresti 2014). The error rates were instead evaluated using a mixed logit model with subject as random effect (Bates et al. 2013). They do not vary significantly with device, UI version, or task necessity. The main effect of goal relevance is significant, but points into the opposite direction with device-oriented subtasks showing lower error rates. There was a significant interaction between necessity and goal relevance (see Table 6.3). Error rates for omissions and intrusions alongside 95%-confidence intervals are given in Fig. 6.5.

6.5.3 Discussion

The largest effect observed in Experiment 2 was the interaction between task necessity and goal relevance. This confirms the theoretical assumption that device-oriented tasks are more prone to errors. At the same time, the significant main effect of goal relevance points into the opposite direction, i.e., device-oriented steps were overall less prone to errors. While this seemingly contradicts the previous finding, it should be interpreted with care. In the presence of an interaction, main effects must not be regarded in isolation. In the current analysis, the interpretation must also take into account that task necessity and goal relevance were unbalanced factors in the design. Most device-oriented tasks were also obligatory (also manifest in the larger error bars for non-obligatory device-oriented steps in Fig. 6.5). Therefore, the main effect of goal relevance should not be interpreted as evidence against the downsides of device-oriented tasks.

In a more general sense, the results are valuable for several reasons. First, they show that human error can be studied well without adding secondary tasks or interrupting the participants during their tasks. Second, the concept of goal relevance has proven beneficial in principle. And finally, the results highlight how little can be predicted based on theoretical concepts alone (here: goal relevance). Only when the application's interaction logic is taken into account (here focused on obligatory versus non-obligatory task steps), differences emerge that are worth further analysis.

This has several implications for the instrumentality of the goal relevance concept for automated usability evaluation. First, the predictive value of the concept that had been shown in Sect. 6.4 needs to be re-established. Second, the dependence on specific properties of the application under evaluation (here: task necessity) highlights the potential benefit of automated evaluation if (and only if) these properties are available

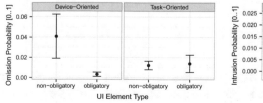 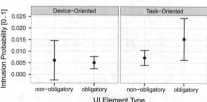

Fig. 6.5 Error probabilities for Experiment 2. Error bars are 95% confidence intervals using the Agresti-Coull method

to the evaluation tools. Whether both can be achieved based on the MBUID models of the kitchen assistant is discussed in the following section.

6.6 Are Obligatory Tasks Remembered More Easily? An Extended Cognitive Model with Cue-Seeking

Confronting the MFG theory of action control from the psychological laboratory with the real-world kitchen assistance application in Experiment 2 has elicited an important UI design aspect that influences human error but is not covered by the theory: task necessity. For device-oriented steps, only non-obligatory tasks show higher error rates (see Fig. 6.5).

How can this be explained by the MFG theory, and how can the cognitive model from Sect. 6.4 account for this result? If the MFG holds, then the lower omission rates for obligatory tasks (e.g., navigating to the next screen) must result from their corresponding goal chunks being higher activated than goal chunks for non-obligatory tasks (e.g., choosing a search attribute). Higher activation can only result from more memory rehearsal or more priming (see Sect. 2.3.3). That obligatory steps are rehearsed more often is highly implausible. Priming on the other hand could be a possible explanation, but what would be the priming source?

The answer proposed here is derived from the concept of embodied cognition introduced in Chap. 2. Humans as embodied and situated beings tend to use environmental cues to reduce cognitive load (Wilson 2002). This led to the addition of a visual *cue-seeking* (or knowledge-in-the-world; Norman 1988) strategy where the user scans the UI for "inviting" elements instead of relying on the internal representation of the current task, only (cf. Gray and Fu 2004). The currently attended UI element then receives *visual priming* that may provide the additional activation of goal chunks for obligatory tasks that the MFG cannot explain.

This proposal of such a strategy also goes in line with Salvucci's criticism of the MFG theory being too focused on memory while neglecting the user's interaction with the environment (Salvucci 2010). In order to examine whether the cue-seeking strategy can explain the empirical findings of Experiment 2, the cognitive model from Sect. 6.4 was extended accordingly.

6.6.1 Model Implementation

As the base model (see Sect. 6.4) only targets the delay effect of device-oriented steps, several changes had to be made. First, the model needed the ability of making omissions and intrusions. This was achieved by using ACT-R's *partial matching* facility. Partial matching mimics human memory imperfections by responding to

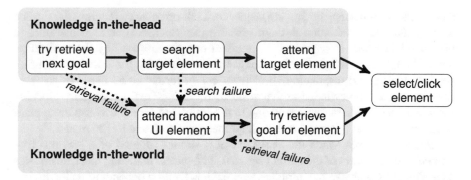

Fig. 6.6 Schematic flow chart of the knowledge-in-the-head and knowledge-in-the-world strategies of the cognitive model. Error-free progress follows the *solid arrows*. *Dotted arrows* denote failures of the annotated types

memory retrievals with similar, but not completely fitting chunks of information. In the current model, this means that requests for the next goal chunk may yield an incorrect result.

Second, the cue-seeking strategy had to be implemented. When pursuing a goal, the model first uses the knowledge-in-the-head strategy of the previous model (see Fig. 6.1), i.e., it tries to follow the memorized step sequence (upper part of Fig. 6.6). Once memory gets weak, the knowledge-in-the-world strategy takes over. Elements on the screen are randomly attended and a memory recall heuristic is used to determine whether this element was part of the current action sequence. If no matching goal chunk is found, the visual search for possible targets is continued (lower part of Fig. 6.6).

Technically, the model[5] is run inside the standard Lisp distribution of ACT-R 6. The ability to interact with the HTML interface of the kitchen assistant is provided by ACT-CV (Halbrügge 2013). The model receives its instructions through ACT-R's auditory system and tries to memorize the necessary steps for the current trial. No specific knowledge about the kitchen assistant is hard-coded into the model.

6.6.2 How Does the Model Predict Errors?

Memory activation (and its noise) is the main explanatory construct used by the model. Omissions are caused by the activation of the respective subgoal being too low. As activation decays over time, omissions are more likely for longer task sequences

---

[5]The source code of the model is available on GitHub (doi:10.5281/zenodo.53198).

and for subgoals that appear late in a sequence. Task-oriented elements of a sequence receive additional activation through priming (see Sect. 6.4) and are therefore in principle less prone to omissions. This effect can be mitigated by the application logic, though. If a subgoal can no longer be retrieved by the knowledge-in-the-head strategy, its corresponding UI element can nevertheless be found by the cue-seeking strategy (knowledge-in-the-world in Fig. 6.6). This is especially probable for mandatory subtasks like navigating to the following screen because these mark situations where no other applicable UI element can be found by the model.

Intrusions happen when the activation of a similar subgoal of a previous trial exceeds the activation of the current subgoal. The partial matching mechanism adds an additional penalty to the intruding subgoal's activation depending on its dissimilarity to the retrieval request.[6] Activation noise is essential for intrusions, but they can also be caused by an old subgoal receiving "misguided" priming from the current goal, e.g., when there is substantial overlap between two consecutive trials. Because task-oriented subgoals receive more priming than device-oriented ones, the model predicts higher intrusion rates for task-oriented UI elements.

Another cause for intrusions is the cue-seeking strategy. ACT-R's activation spreading mechanism allows priming from the current focus of the visual system to declarative memory elements, comparable to the horizontal triggers of the 'contention scheduling' model (Cooper and Shallice 2000, see Sect. 2.3.1). When the model searches the screen for "inviting" elements (*attend-random-ui-element* in Fig. 6.6), the currently attended element primes subgoals that correspond to it regardless of whether they belong to the current goal or a previous one.

6.6.3   Model Fit

The model predictions are sensitive to several global ACT-R parameters that affect activation. Activation decay (bll) was kept at the default of 0.5, activation noise (ans) was varied between 0.5 and 0.6, priming (mas) was set to 3.5 and the partial matching penalty (mp) was varied between 4.0 and 4.5.

The model was run 1000 times and all errors made were collected (dotted lines in Fig. 6.7). The quantitative fit as computed based on the error probability for each combination of omissions and intrusions and the four UI element types defined above is promising with $R^2 = 0.915$ and RMSE = 0.003. Qualitatively, intrusion rates are matched very well. Omissions are matched well for device-oriented, but overestimated for obligatory task-oriented UI elements.

---

[6]The dissimilarities are computed by ACT-R based on the number of mismatching information units, here trial and current subgoal. No user-specified similarity function is used by this model.

6.6.4   Discussion

The initial cognitive model from Sect. 6.4 was extended with a cue-seeking strategy that the model applies after facing memory retrieval failure. While the overall goodness-of-fit of the model is good and especially the case of non-obligatory device-oriented tasks showing the highest error rate is reproduced well (see Fig. 6.7), there are several limitations. The sheer number of mechanisms used (decay, priming, etc.) leads to a complex cognitive model that is rather sensitive to changes of the respective ACT-R parameters. Future research will show whether this affects the generalizability of the model. The same holds for the empirical basis of the model and the representativity of the tasks used during the experiment. Finally, the proposed cue-seeking strategy probably oversimplifies human visual search behavior.

The model nevertheless provides several improvements when compared to existing MFG models of procedural error. Most important, while Altmann and Trafton (2002) discuss how the environment can provide cues that prime pending goals, the model presented here is the first one that actively uses this strategy. As a by-product, this leads to the prediction of intrusions, an important error class that is often not well captured (e.g., Trafton et al. 2011; Li et al. 2008; Hiatt and Trafton 2015).

With respect to the overall goal of this work, i.e., analyzing the applicability of automated usability evaluation for model-based applications, the current results present a significant step forward. With goal relevance and task necessity, two factors influencing human error have been validated (see Fig. 6.8). As both factors can be captured by the cognitive user model presented here, this at least theoretically allows automated evaluation.

On the down side, the model has been created post-hoc, i.e., after data collection, and its implementation has been informed by the empirical results. Whether the model captures general processes of human action control or just the very specifics

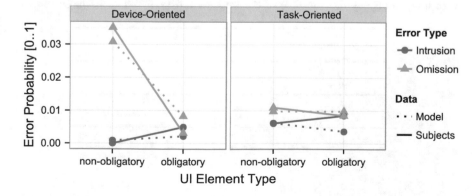

Fig. 6.7   Model predictions and empirical error rates

of Experiment 2 can only be decided based on new evidence. This will be provided in the following sections.

6.7   Confirming the Cue-Seeking Strategy with Eye-Tracking (Experiment 3)

While the theoretical model presented in Sect. 6.6 is targeting only error rates, the addition of the cue-seeking strategy predicts behavior changes in other areas as well that allow empirical validation. In the visual domain, using the bottom-up cue-seeking strategy means continuously searching the UI for any suitable element, while the top-down memory-based strategy just looks for the specific element that is needed to perform the current task. Therefore, an eye-tracking experiment was designed that a) should reproduce the previous findings and b) should confirm the predictions of the knowledge-in-the-world assumption in the visual domain as well.

6.7.1   Methods

Participants

24 members of the Technische Universität Berlin paid participant pool, 15 women and 9 men, aged between 18 and 54 (M = 29.4, SD = 9.4), took part in the experiment conducted in January 2015. As the instructions were given in German, only fluent German speakers were allowed. Written consent was obtained from all participants.

Materials

A personal computer with 23″ (58.4 cm) touch screen and a 10″ (25.7 cm) tablet were used to display the interface of the kitchen assistant. While the large screen

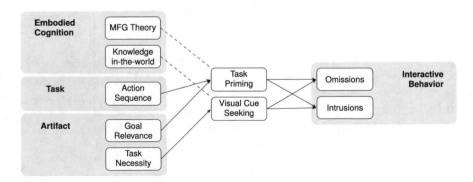

Fig. 6.8 Human error as a function of goal relevance and task necessity

operated in landscape mode, portrait mode was used for the tablet. Both variations of the kitchen assistant's UI from Experiment 2 were used (see Sect. 6.5). All user actions were recorded by the computer system and the participants' performance was additionally recorded on videotape for subsequent error identification.

The participants' gaze was recorded using a SR Research Ltd EyeLink II head-mounted eye-tracker. Only the dominant eye was tracked; the sampling frequency was 250 Hz. Gaze recording was only applied during task blocks that used the large monitor because the visual angles between UI elements on the tablet were too small to obtain an unambiguous mapping from gaze to element. Nevertheless, the eye-tracker was worn by the participants during the complete procedure as its scene camera view was recorded for subsequent error identification. The experiment was conducted in a laboratory with fixed lighting conditions.

Design

The experiment features a four-factor within-subject design, the factors being the physical device used, the UI variant, whether a sub-task was obligatory, and whether it was device-oriented as opposed to task-oriented. Dependent variables were user errors, task completion times, and gaze position. The participants completed a total of 46 tasks grouped into 4 blocks. Physical device, UI variant, and block sequence were varied randomly, but counterbalanced across the experiment.

Procedure

After having played a simple game on each of the two devices to get accustomed with the respective touch technology, the participants received training on the kitchen assistant. The training covered all parts of the application that were used during the actual experiment. Each block of tasks began with relatively simple tasks like "Search for German main dishes and select lamb chops". Afterwards, the ingredients to a recipe were collected and some of them were added to a shopping list that is part of the kitchen assistant (e.g., "Create a shopping list for six servings and check-off garlic"). The instructions followed the experiment described in Sect. 6.5 closely, with only one new type of device-oriented non-obligatory tasks added in order to gain more insights about this special combination.[7] When a participant made an error during a trial, they were not interrupted but informed after the trial and were given the chance to repeat this trial a single time. The complete procedure lasted approximately 45 min.

6.7.2 Results

Errors

A total of 6921 clicks were recorded. There were 85 (1.2%) omissions and 53 (0.8%) intrusions. On trial level, 12.4% of the user tasks were not completed correctly on

---

[7]The full instructions are available at Zenodo (doi:10.5281/zenodo.268596).

Table 6.4 Mixed logit model results for Experiment 3. OR: Odds ratio for errors

Factor name	OR	95% CI of OR		z	p
Device (Tablet vs. Screen)	1.06	0.75	1.52	0.34	.730
UI variant (Simple vs. Complex)	0.95	0.67	1.36	−0.27	.789
Device-oriented step	0.94	0.43	2.37	−0.14	.886
Non-obligatory step	2.10	1.02	5.09	1.84	.066[†]
Interaction Dev.-or. × Non-obligatory	7.94	2.91	19.44	4.33	<.001**

Note **highly significant ($p < .01$); *significant ($p < .05$); [†] marginally significant ($p < .10$)

the first try. Because of the unequal number of observations per trial, mixed models were used to analyze the data (Bates et al. 2013). Neither physical device nor UI variant showed relations to the error rate (mixed logit model with subject as random factor, see Table 6.4 Fig. 6.9).

Eye-Tracking

The gaze of three participants could not be recorded because of calibration failure. For the remaining 21 participants, the raw screen coordinates were mapped to dynamic areas of interest (AOI) around the UI elements using a computational bridge between the eye-tracker and the web browser that rendered the UI (Halbrügge 2015a). Individual gaze positions were collapsed into fixations using a hidden Markov model (HMM) approach. Based on common rule-of-thumb values of maximum 100 deg/s for fixations and minimum 300 deg/s for saccades (Salvucci and Goldberg 2000), the parameters for the states of the HMM and the respective transition probabilities were estimated from the recorded data.

In order to examine the knowledge-in-the-world assumption, the gaze recordings were split into segments based on the clicks within a trial (e.g., a trial with five clicks yielded four segments). Because of the varying length of these segments, fixation rate (number of fixations divided by segment length) is used as dependent variable. A mixed model with subject and sub-task as random factors yielded significantly higher

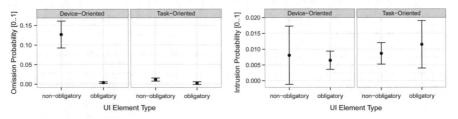

Fig. 6.9 Error probabilities for Experiment 3. Error bars are 95% confidence intervals using the Agresti-Coull method

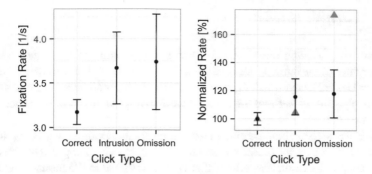

Fig. 6.10 Fixation rates for Experiment 3. Error bars are 95% confidence intervals

rates for segments before erroneous clicks compared to correct clicks ($F_{2,1952.5} = 5.3$, $p = .005$, see Fig. 6.10). This could still mean that the users were recurrently fixating the ultimately clicked UI element instead of searching the screen. This was addressed by counting the fixations on the AOI corresponding to the element that concluded its segment. Because the resulting counts are extremely right-skewed (median $= 1$), they were divided into two groups for statistical analysis. A logit mixed model with subject and sub-task as random factors revealed that the probability to fixate the ultimately clicked element at least once during a segment was actually lower before errors ($z = -2.5$, $p = .012$).

6.7.3 Results Discussion

The results partially reproduce the findings of the previous studies. Compared to Experiment 2 (Sect. 6.5), the combination of non-obligatory device-oriented tasks yielded a much higher error rate. This is probably due to the changed set of user tasks that seems to have included device-oriented task steps that were much harder to remember than the ones used before.

The eye-tracking data confirms the knowledge-in-the-world assumption nicely. The higher fixation rate before erroneous clicks matches the proposed process of (random) search for 'inviting' elements on the surface of the UI. In principle, this could also be caused by memory-based processing, e.g., a re-fixation strategy to strengthen the activation of the current subgoal through visual priming. But the AOI-based analysis shows that the frequent fixations before errors are on other elements than the one that is eventually clicked.

6.7.4 Cognitive Model

The new data presented here allows constraining the existing model. The model fit to the error rates of the current data is still good with $R^2 = 0.76$, but somewhat increased RMSE = 0.044. The eye-tracking results confirm the theoretical assumptions at least qualitatively. But what about the actual quantitative predictions of the model in the visual domain?

As ACT-R's visual module operates on an abstract attention layer (as opposed to raw eye movements), the model's visual behavior cannot be compared directly to the human sample. Nevertheless, effects that were significant in the human data should be present in the model results as well. The model from Sect. 6.6 produces fixation rates that are actually lower for erroneous trials compared to correct trials, in contrary to both the theory and the human data. This happens because the memory test that is part of the knowledge-in-the-world strategy ("try retrieve goal for element" in Fig. 6.6) is not restricted in time. If it fails to retrieve a matching goal chunk, this is only signaled after approximately one second. As a result, the visual search process is slowed down considerably.

Recent neuroimaging studies (Anderson et al. 2016; Borst et al. 2016) based on the famous fan effect (Anderson and Reder 1999) propose that the familiarity of a cue is determined by a different process than associative retrieval and that this process is very fast (a few hundred milliseconds). As ACT-R in its current version does not incorporate such a familiarity process, the bad fit of the original model to the gaze data is understandable. How can this be improved without re-implementing ACT-R's memory system?

Revised Model
In order to speed up the knowledge-in-the-world strategy, ACT-R's temporal buffer (Taatgen et al. 2007) is used. A timer is started with the initiation of the memory test and a single production was added that abandons the retrieval after a fixed amount of time ticks. Based on the observed fixation rate between 3 and 4 per second before erroneous clicks, the tick threshold was set to approximately 250 ms (9 to 12 ticks using ACT-R standard parameters).[8]

Goodness of Fit
Comparing the model predictions based on the fixation rates is tricky because ACT-R only models shifts of attention and assumes completely stable gaze otherwise, which results in unnaturally low fixation rates. An exploratory approach to this based on the relative change compared to the 'correct click' baseline is shown in Fig. 6.10 (right). The corresponding fit is unimpressive with $R^2 = 0.41$, but should be interpreted with care. Of higher importance is the qualitative result that error trials show increased fixation rates.

---

[8]The source code is available on GitHub (doi:10.5281/zenodo.55223).

After having optimized the model for visual behavior, the fit in the error domain degrades slightly with $R^2 = 0.70$ and RMSE $= 0.044$.

6.7.5 Discussion

Eye-tracking has been used before to predict special subtypes of procedural error (Ratwani and Trafton 2011) with good success. Ratwani's system uses a cognitive model based on the MFG theory in combination with a statistical classifier that uses eye-tracking data as input to predict and prevent postcompletion error. The approach presented here differs therefrom in two important ways. First, it targets not only omissions of the final step within a sequence, but other types of omissions and intrusions as well. And second, the model presented here tries not only to predict *when* a user makes an error, but also *why* this happens, which led to the incorporation of the necessary visual processes into the model.

While providing good fits to the data, the model also has several limitations. First, the model only covers expert behavior. The initial formation of the task sequence by novice human users is beyond its capabilities. The model also does not account for errors caused by the UI design violating general expectations of its users towards computer systems.

Second, the empirical basis of the model is limited. So far, data has only been collected using a single paradigm. Changing the experimental tasks resulted in changed error rates, but the overall pattern remained stable and the model fit is still satisfactory. While the current addition of eye-tracking data represents a significant extension of its empirical foundation, the generalizability of the model still needs further investigation.

6.8 Validation in a Different Context: Additional Memory Strain Through a Secondary Task (Experiment 4)

In order to test not only the reliability of the effects found in Experiments 2 and 3, but also their generalizability and the validity of the model, the experimental paradigm was substantially changed. As discussed above (see Sect. 6.5), exposing the participants of a study to secondary tasks and interruptions leads to increased error rates, but lowers the ecological validity of the experimental results. On the other hand, dual-task paradigms provide the opportunity to challenge the user model by comparing its predictions to empirical data collected in such a scenario. Finally, more direct comparisons to other research are possible using this paradigm (e.g., Byrne and Bovair 1997; Ament et al. 2010; Trafton et al. 2011).

Fig. 6.11 Examples of the pictograms used in Experiment 4. Image credits: Baby © UN OCHA, CC-BY 3.0; Door, Cup with tea bag © Freepik, CC-BY 3.0

For these reasons, a dual-task paradigm using the kitchen assistant (see Sect. 3.2) and another memory-intense task was designed. A working memory updating (WMU; Ecker et al. 2010) task was chosen as secondary task because similar tasks have been used before in this scenario (Ament et al. 2010) and because of the availability of a validated cognitive model of this task that could be combined with the cognitive model from the previous Sect. 6.7 (Russwinkel et al. 2011).

The WMU task should interfere with the primary kitchen assistance task by a) periodically interrupting the user and b) additional memory strain. Based on previous results (e.g., Byrne and Bovair 1997), this should result in increased error rates in the dual-task condition compared to a single-task baseline.

6.8.1 Method

Participants
Twelve members of the Technische Universität Berlin paid participant pool took part in the experiment. There were four men and eight women, with their age ranging from 18 to 51 (M = 33.7, SD = 9.5). As the instructions were given in German, only fluent German speakers were allowed to take part. Written consent was obtained from all participants.

Materials
The experiment was conducted in a neutral laboratory. A personal computer with 23″ (58.4 cm) monitor with optical sensor 'touch' technology was used to display the interface of the kitchen assistant. Seven pictograms of common household interruptions (e.g., phone ringing, doorbell, baby crying; see Fig. 6.11) served as stimuli of the WMU task. The stimuli were superimposed on the UI of the kitchen assistant using dedicated Javascript code running within the browser that displayed the assistant. All user actions were recorded by the computer system. The participants' performance was additionally recorded on videotape for subsequent error identification.

Recipe Task   Present   Perform   Ask   WMU
Instruction   WMU Target   Recipe Task   WMU Count   Feedback

Fig. 6.12 Sequence of screens within a single trial in the dual task condition

Design

A three-factor within-subjects design as applied, the factors being goal relevance (device- versus task-orientation), task necessity (non-obligatory vs. obligatory), and secondary task difficulty (none versus onset to onset stimulus intervals 5 s, 4 s, 3 s). User tasks were grouped into four blocks of eleven to twelve individual tasks. Each participant was randomly assigned to one of eight pre-selected block sequences so that block position and block succession were counterbalanced across participants as well. The secondary WMU task was always introduced after the completion of the first block and its sequential demand was gradually increased from 5 s stimulus interval in the second to 3 s in the fourth and last block. Each interval was split into equally long stimulus and blank phases.

Procedure

Every block started with comparatively easy recipe search tasks, e.g., "search for German main dishes and select lamb chops".[9] Users would then have to change the search attributes, e.g., "change the dish from appetizer to dessert and select baked apples". The second half of each block was made of more complex tasks that were spread over more screens of the interface and/or needed memorizing more items. The participants either had to create ingredients lists for a given number of servings, or had to make shopping lists using a subset of the ingredients list, e.g., without salt and flour. All instructions were read to the participants by the experimenter. Every individual trial was closed by a simple question the participants had to answer to keep them focused on the kitchen setting, e.g., "how long does the preparation take?" During each instruction phase the complete screen was blanked (see Fig. 6.12).

In the dual-task condition, one of the seven WMU stimuli was selected as target for the current trial and presented to the participants after the instructions for the next trial had been given. Subsequently, the UI of the kitchen assistant was uncovered. WMU stimuli appeared in random order on the lower right of the screen and the participants had to count the number of appearances of the target stimulus. After the completion of the trial, the screen was blanked again and the participants were asked how often they had seen the WMU target. With an initial training phase and exit questions the whole procedure took approximately one hour.

---

[9]English translations of the actual instructions are given here for reasons of comprehensibility. The original instructions are available at Zenodo (doi:10.5281/zenodo.268596).

Table 6.5 Mixed logit model results for Experiment 4. OR: Odds ratio for errors

Error class	Factor name	OR	95% CI of OR		z	p
Omissions						
	With WMU task	0.97	0.59	1.64	−0.12	.906
	Device-oriented step	0.55	0.19	1.67	−1.12	.264
	Non-obligatory step	3.18	1.46	8.36	2.65	.008**
	Interaction Dev.-or. × Non-obligatory	2.42	0.58	9.34	1.26	.208
Intrusions						
	With WMU task	0.85	0.57	1.29	−0.81	.416
	Device-oriented step	0.94	0.39	2.62	−0.14	.892
	Non-obligatory step	5.51	2.59	14.37	4.13	<0.000**
	Interaction Dev.-or. × Non-obligatory	0.11	0.01	0.68	−2.06	.040*
All procedural errors						
	With WMU task	0.89	0.65	1.24	−0.72	.470
	Device-oriented step	0.73	0.37	1.52	−0.90	.369
	Non-obligatory step	4.58	2.61	8.85	5.03	<0.000**
	Interaction Dev.-or. × Non-obligatory	0.75	0.26	2.03	−0.55	.582

Note **highly significant ($p < .01$); *significant ($p < .05$); [†]marginally significant ($p < .10$)

6.8.2   Results

A total of 3464 user actions and 407 min of video were recorded. The system logs were synchronized with the videos and semi-automatically annotated using ELAN (Wittenburg et al. 2006).

Errors
There were 88 (2.5%) omissions and 133 (3.8%) intrusions. On trial level, 29.5% of the kitchen assistant tasks were not completed correctly by the participants on the first try. Contrarily to the assumptions given above, the error rate did not increase in the dual-task condition, nor when interruptions by the secondary task became more frequent. Results of mixed logit analyses with subject and trial as random factors are given in Table 6.5. Error probabilities and confidence intervals are shown in Fig. 6.13.

There was a significant influence of the WMU task on the time needed to perform the recipe task. In the dual-task condition, participants needed approximately 100 ms longer per individual click (mixed model with click type as defined in Sect. 5.3 and subject as random factors, $t_{2695} = 2.31$, $p = .021$).

Working Memory Updating Task
Contrarily to the assumptions, the rate of errors in the memory updating task did not increase with the demand of task. Adding the block to a mixed logit model with task block and subject as random factors did not explain more variance ($\chi_2 = 0.12$, $p = .942$). Descriptively, there is a small increase, but the error rate is already very high (above 0.3) in the easiest condition as shown in Fig. 6.14.

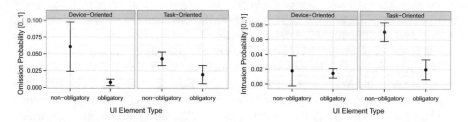

Fig. 6.13 Error probabilities for the main recipe task. Error bars denote 95% confidence intervals using the Agresti-Coull method

6.8.3 Results Discussion

Interruptions are often used in error research as a means to increase the error base rates (e.g., Trafton et al. 2011; Li et al. 2008). In line with this thinking and based on the results of previous research (e.g., Byrne and Bovair 1997; Ament et al. 2010), it was expected that the increased memory load due to the WMU task would result in degraded performance in the main recipe task. But the empirical data tells a different story. While the participants needed more time to complete the recipe tasks, they did not make significantly more errors. Why is this the case?

First, the baseline error rate of 6.4% is already quite high, in particular compared to the 2.0% observed during Experiment 3 (Sect. 6.7) and the 1.6% observed during Experiment 2 (Sect. 6.5). This could be due to the blanking of the screen during the instruction phase that was added to the procedure in the present experiment. The blanking should have impaired the learning of the UI of the kitchen assistant. In previous studies, the participants could visually plan their actions during the instruction phase, while the current experiment demanded memorizing all subgoals without any visual reference.

Fig. 6.14 Error Probabilities per Block for the WMU Task. Error bars denote 95% confidence intervals based on the Agresti-Coull method

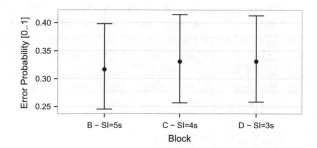

Second, the high error rate of the WMU task suggests that the participants spent most attention on the recipe task. But as the WMU performance remained rather stable with increased difficulty, this point remains unsatisfactory.[10] The analysis of the tasks based on the cognitive model presented below will provide additional insights.

The prolongation of the time needed to perform the recipe task in the dual-task condition is in line with the expected interference between both tasks. The real effect is probably underestimated by the analysis, because learning effects of approximately -50 ms per block were observed in Experiment 1 (Table 6.1). Assuming that learning still took place in the current study, it should have counteracted the prolongation effects of increasing WMU complexity.

6.8.4 Cognitive Model

The MFG theory qualitatively predicts that having to perform two memory-intensive tasks at the same time leads to more errors, but the data does not confirm this. Does this disprove the theory? In order to elaborate on this, the MFG-based cognitive model from the previous Sect. 6.7 was combined with an existing model of WMU (Russwinkel et al. 2011) using the threaded cognition extension of ACT-R (Anderson et al. 2004). The threaded cognition theory (Salvucci and Taatgen 2008) assumes that task switching is not necessarily conscious behavior, but may emerge as concurrent tasks have to wait for cognitive resources (e.g., memory, vision) that are currently held by other tasks.

For the current experiment, the model from Sect. 6.7.4 was adapted to threaded cognition by adding extra checks for the current availability of the declarative module to several productions. All numerical ACT-R parameters remained unchanged.

The pictogram counting part was adapted from an existing model that has been tested in different kind of tasks and settings (Russwinkel et al. 2011). The WMU model uses a single representation (i.e., memory chunk) for each target that pairs it with its current count. In the combined model, the ACT-R mechanism of 'buffer stuffing' is used to detect the visual targets of the WMU task, i.e., no active search is performed, but a general vigilance system is used that notifies the model after appearance of a new object in the visual field. In case this new object is found at the right bottom of the screen and both the visual and the declarative modules are available, the model attends the object and at the same time retrieves the most highly activated WMU count chunk. Because of activation noise, the declarative module may return an older copy of that chunk which subsequently leads to an error.

---

[10]Unfortunately, Ament et al. (2010) give no results of the secondary task that could be used for comparison.

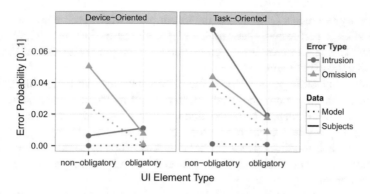

Fig. 6.15 Model predictions and empirical error rates (Experiment 4)

Goodness of Fit

The combined model[11] was run 500 times and all resulting errors and completion times were recorded. Contrarily to the expectations, but consistent with the empirical findings, the combined model does not show an increased error rate when the WMU task is present (OR = 1.04, well within the empirical 95% CI from 0.62–1.20).

Regarding the effects of device-orientation and task necessity, the overall model fit is not good with $R^2 = 0.174$ and RMSE = 0.029. This is mainly due to the unexpectedly high intrusion rate for non-obligatory task-oriented steps (see Fig. 6.15). When regarding only omissions, the fit is much better with $R^2 = 0.789$ and RMSE = 0.014.

Interestingly, the high omission rate for non-obligatory task-oriented steps found empirically is matched by the model (compare to Fig. 6.7). This unexpected results needs further research.

The combined model does show longer click times in the dual-task condition. Here, each click takes 120 ms longer on average, which is close to the empirical effect of 94 ms (95% CI from 14–174 ms).

6.8.5   Discussion

When the combined model performs both the recipe and the WMU task, the number of errors in the recipe task does not increase, it just takes longer to perform the individual actions. This means that the empirical results, although being unexpected, fit with the theoretical underpinnings presented above. Close inspection of the model traces shows that the model predictions are caused by both tasks demanding visual

---

[11]The source code of the combined model is available for download on GitHub (doi:10.5281/zenodo.55224).

and memory resources. Only during the motor phase of the recipe task model (i.e., when a click is performed, "select/click element" in Fig. 6.6), the WMU task can take over. This observation is also consistent with the high error rate of the WMU task during the experiment.

From the scientific point of view, the results uncover two major limitations of the model. First, the increased intrusion rate is not matched. At the same time, a proper theoretical explanation for the high number of intrusions in the current study is still missing. Further research is needed. Second, the stark behavior change even in the single task condition needs further attention. The environment seems to play an important role while listening to the instructions and planning the corresponding action sequence.

With regard to the overall goal of this work, i.e., automated usability evaluation for model-based applications, other aspects get into the focus. First, while the ecological validity of the study is smaller compared to the previous experiments 2 and 3, the fit of the model to the empirical omission rates is still good with $R^2 = 0.789$. Second, the poor prediction of intrusions in this study poses a problem to the model only if the observed intrusion pattern a) is valid across different contexts and b) is a function of the UI design. While a) could only be answered with new data, b) can probably be rejected based on the results of Gray (2000) reported in Sect. 2.3.2 and based on the MFG theory. In both views, intrusions are mainly a result of competition with 'memory clutter' from previous trials.

6.9   Chapter Discussion

How important are errors?
Base rates are low. Across three experiments, there was an overall omission rate of 1.4% and intrusion rate of 1.3%. In the case of intrusions, Experiment 4 yielded a much higher rate than the previous studies with 3.8% compared to an average of 0.6%. Nevertheless, these low base rates led to unsuccessful task completion in 9.5% (Experiment 1) to 29.5% (Experiment 3) of the trials.

It is important to note that other, non-procedural types of errors were virtually absent in the experimental data. Neither knowledge-based mistakes (located on the top of the step-ladder model of Sect. 2.1), nor sensory-motor slips (at the bottom of the step-ladder) were observed in noteworthy quantities.[12] The focus on procedural error in this work that has been set very early in Chap. 2 can therefore be justified based on the empirical data as well.

---

[12]It is true though that some intrusions could also be regarded as motor slips as these can not be distinguished from intrusions on the phenotypical level.

Table 6.6 Mixed logit model results for Experiment 2–4. OR: Odds ratio for errors

Error class	Factor name	OR	95% CI of OR		z	p
Omissions						
	Device-oriented step	0.47	0.24	0.93	−2.28	.023*
	Non-obligatory step	2.38	1.42	4.34	3.10	.002**
	Interaction Dev.-or. × Non-obligatory	10.24	4.82	21.07	6.29	<.001**
Intrusions						
	Device-oriented step	1.02	0.56	1.97	0.06	.954
	Non-obligatory step	2.99	1.75	5.57	3.84	<.001**
	Interaction Dev.-or. × Non-obligatory	0.12	0.02	0.43	−2.84	.005**
All procedural errors						
	Device-oriented step	0.73	0.47	1.16	−1.42	.157
	Non-obligatory step	2.72	1.86	4.16	4.95	<.001**
	Interaction Dev.-or. × Non-obligatory	3.17	1.84	5.35	4.29	<.001**

Note **highly significant (p < .01); *significant (p < .05); †marginally significant (p < .10)

How important are goal relevance and task necessity?
Table 6.6 shows results from a combined analysis of the data collected during the three experiments reported in this chapter.

The strongest effect overall is the interaction between task necessity and goal relevance. Non-obligatory device-oriented steps are most prone to omissions. The OR of 10.24 means that—assuming a base omission rate of 1%—omissions are more than nine times more likely for this specific combination.[13]

The results for intrusions are mixed. Based on the MFG theory, higher intrusions rates would be expected for task-oriented goals. This could not be confirmed. The base intrusion rate is much lower than the omission rate, though, making analysis harder. The strongest effect in the intrusion domain is that intrusions of non-obligatory device-oriented steps are much less frequent (OR = 0.12). This stems mainly from Experiment 4, the other experiments showed a different pattern. Further research is needed to gain more evidence.

Overall, procedural errors happen more often during non-obligatory steps, this effect is especially strong if these are also device-oriented.

How does this relate to AUE for model-based applications?
For the purpose of automated usability evaluation, the empirical findings imply that both the device-orientation and the task necessity factors should be accessible to a potential AUE system. Can both factors be provided by the MBUID models that are accessible in runtime frameworks like the MASP (see Sect. 3.2)?

Device-orientation can be deduced statically from the AUI and the task model as the AUI differs between interactive UI elements that allow to transmit information

---

[13]calculation: relative risk $= \frac{OR}{(1-P_{baseline})+(P_{baseline}\cdot OR)} = \frac{10.24}{0.99+0.1024} \approx 9.37.$

(type 'input') and elements that do not (type 'command'). This matches the main property of device-oriented tasks that they do not convey any information related to the users' goals (see introduction of the concept in Sect. 6.1 and a more detailed discussion in Sect. 8.1). Examples for device-oriented tasks together with their MBUID meta-information are given in Fig. 6.4.

Task necessity on the other hand is a property that—while it could be modeled using CTT notation—is often only introduced at the level of the final UI. This means that it can not be deduced from a development time model. An executable user model that interacts with the final UI is necessary, instead.

6.10 Conclusion

The current chapter presented how UI meta-information can be used to predict the effectiveness (in terms of task success) of a given UI and set of tasks. The basis for the prediction is a new theoretical model of sequential action control and error. Empirical data has shown that existing activation-based models (e.g., Trafton et al. 2011; Tamborello and Trafton 2015; Rukšėnas et al. 2014) do not capture important aspects of real-world applications, namely that some tasks are obligatory (e.g., navigating to a subsequent page) while others are not (e.g., selecting an option field). The data (and daily life) clearly show that obligatory tasks are less prone to omissions, but the MFG (Altmann and Trafton 2002, see Sect. 2.3.3) and also the GUM (Butterworth et al. 2000, see Sect. 4.3.3) lack a mechanism that can explain this effect. The model proposed here closes this gap by introducing a second *knowledge in-the-world* (Norman 1988) strategy which incorporates not only susceptibility to external cues (e.g., 'bottom-up' processing, or horizontal lines in Cooper and Shallice 2000), but active visual search on the UI for elements that fit to the current goal.

A second factor that determines the error proneness of a task is device-orientation (Ament et al. 2013; Gray 2000). Such tasks differ from their task-oriented counterparts by not contributing to the user's overall goal (at least not directly). They are called device-oriented because they are somewhat forced on the user by the device. Empirically, device-orientation is connected to longer task completion times and higher omission rates. In the cognitive model, this is explained by a lack of priming from the user's goal to the subgoals that correspond to the atomic tasks that constitute an action sequence.

Based on task necessity and device-orientation, the model explains between 17 and 92% of the variance in the error data from the experiments presented in this chapter. How can this be utilized for automated usability evaluation of MBUID systems? The ACT-R model presented here does not directly interact with the MASP system, but with an automatically created mock-up (Halbrügge 2015b). This allows fast-time simulation, but hinders direct access to the meta-information provided by the MASP.

It is worth noting that both task necessity and goal relevance can be derived from runtime information of the MASP system. How this can be done will be shown in Chap. 8.

The next chapter elaborates on a major weakness of the model, namely that it treats users as unknowing beings without any expectations or preknowledge. The effects of preknowledge on the empirical results from the current chapter are examined and the model is extended with external sources of knowledge to account for these effects.

Chapter 7
The Competent User: How Prior Knowledge Shapes Performance and Errors

In this chapter [1]:

- Unfamiliarity with concepts used in an interface may cause errors.
- How can this be captured by a predictive system?
- Modeling concept familiarity needs knowledge sources external to the UI.

System developers have often a very specific view on how an application should be used. Some authors have even highlighted that 'typical' developers find joy in situations that are rather avoided by most users (e.g., challenging environments with high complexity; Tognazzini 1992).

Especially in the context of using development time models for automated usability evaluation, this leads to the question of the *validity* of the task models generated by developers. While the internal consistency of the models is more or less perfect, i.e., the AUI, CUI, and FUI levels fit perfectly together and fit perfectly to the initial task model, the initial task model might contain implicit assumptions about the tasks and the users that might be untrue. In other words: the mental model [2] of the developers may not match the mental model of the users of the system. This kind of mismatch can lead to usability problems that are only uncovered very late in the development process, which makes fixing them very costly (Nielsen 1993).

---

[1] Parts of Sect. 7.1 have already been published in Halbrügge et al. (2015a), parts of Sect. 7.2 have already been published in Halbrügge and Schultheis (2016).

[2] Note that 'mental model' refers mainly to the *procedural* aspect of handling a system, here. This kind of mental model is located on the rule-based level of the step-ladder model in Fig. 2.2. In other contexts, the term mental model may rather refer to declarative knowledge and expectations, e.g., the expectation that hitting 'save' leads to a persistent state that one can get back to even if the system has been powered off in between. This is *not* the notion of 'mental model' used in this work.

© Springer International Publishing AG 2018

M. Halbrügge, *Predicting User Performance and Errors*, T-Labs Series
in Telecommunication Services, DOI 10.1007/978-3-319-60369-8_7

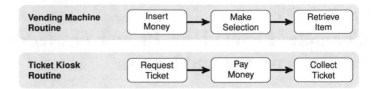

Fig. 7.1 Example of stored procedure mismatch from Baber and Stanton (1996). Problems arise when the system model of a machine strictly follows the 'Ticket Kiosk' scheme and users apply their 'Vending Machine' routine. According to Baber and Stanton, such machines often carry notes that say "Do not put your money in first"

An informative example for this kind of mental model mismatch has been reported by Baber and Stanton (1996). According to Baber and Stanton, users commonly apply one of two routines for buying stuff (see Fig. 7.1). They call these the 'vending machine routine' (insert money, make selection, retrieve item) and the 'ticket kiosk routine' (request ticket, pay money, collect ticket). Problems arise if the developers strictly followed the 'ticket kiosk routine'. In the worst case, the resulting system treats putting in money without a selection as user error and displays nothing but a cryptic error code to communicate this. Potential customers may interpret this system reaction as system failure and may simply give up trying to buy items from the machine. Baber and Stanton observed that such machines often carry stickers that say "Do not put your money in first".

It should be obvious that vending machines should not need additional stickers to correct their poor usability. Instead, the development process should ensure that such problems are fixed before a system is rolled out to the public. This chapter explores whether and how the user modeling approach can be used to spot mental model or concept mismatch. The problem is approached empirically by analyzing the effects of the concepts used in an interface on the behavior of the users of that interface. While the integration of external knowledge with the user model yields important findings, the work presented in this chapter remains preliminary. Future research will show whether the kind of mismatch shown in Fig. 7.1 could be spotted by the modeling approach.

7.1   The Effect of Concept Priming on Performance and Errors

The following analysis serves two goals. First, it aims to quantify the effects of preknowledge on user behavior—how important is it? Second, the cognitive model that has been developed in the previous chapter should be able to capture such effects. The analysis is therefore also a means to check the validity of the theoretical assumptions of the model (e.g., the roles of activation and priming).

The first question is approached by a re-analysis of the combined data[3] from the Experiments 2 to 4. It is important to note that this analysis is orthogonal to the analyses from the previous Chap. 6, i.e., no alpha adjustment needs to be performed during statistical analyses (Bortz 1999). The new analysis is concerned with how long human users need to perform simple tasks with the kitchen assistant, how often they make errors, and which UI elements are tied to these errors.

The second question, i.e., whether the cognitive model can capture the effects of preknowledge on user behavior, it approached by augmenting the model with knowledge about the world based on Wikipedia content. This is following Salvucci's work on the integration of DBpedia (Lehmann et al. 2015) into ACT-R (Salvucci 2014). The modeling effort presented in this section relies mainly on the general relevance of different Wikipedia articles. The higher the number of links inside Wikipedia that point towards an article, the higher the relevance of the article and the entity or concept that it explains. The data suggests that UI elements that correspond to highly relevant concepts are handled better than elements that correspond to less relevant concepts.

The basic assumption is that subgoals receive priming from the general concepts that they represent. Hitting a button labeled "Search" is connected to the concept of search; choosing an option called "Landscape" in a printing dialog is related to the concept of landscape. If the general concept that is semantically linked to a subgoal is highly activated in the knowledge base, the respective subgoal should receive more priming, resulting in a higher overall activation of the subgoal. Taken together with the MFG, this results in three high-level predictions for subgoals, corresponding UI elements, and their respective concepts:

1. The task completion time should decrease with concept activation.
2. The omission rate should decrease with concept activation.
3. The intrusion rate should increase with concept activation.

7.1.1   Method

In order to assess the three ontology-based predictions stated above, a re-analysis of the data from Experiment 2–4 is performed. The focus of the analysis is on a single screen of the kitchen assistant that allows searching for recipes based on predefined attributes. A screenshot of the search attribute form translated to English is given in Fig. 3.2. The search attributes are grouped into nationality (French, German, Italian, Chinese) and type-of-dish (Main Course, Pastry, Dessert, Appetizer). Three health-related search options were excluded from the analysis as they were neither well represented in the experimental design, nor in the ontology. For the eight remaining

---

[3]The ideas presented in this section have first been published in Halbrügge et al. (2015a). At that time, only the data from Experiment 2 was available, which left some effects insignificant. The incorporation of the data of the following two experiments reinforces the overall picture from the previous analysis and allows some further insights.

Table 7.1 Semantic mapping between UI and ontology. Inlink count obtained from DBpedia 3.9 (Lehmann et al. 2015). Subtitle-based word frequency (per 10^6 words) from Brysbaert et al. (2011)

Concept	UI label	DBpedia entry	Inlink count	per 10^6 links	Word freq.
German	Deutsch	Deutschland	113621	2474.3	10.2
Italian	Italienisch	Italien	56105	1221.8	6.2
Chinese	Chinesisch	China	10115	220.3	8.2
French	Französisch	Frankreich	79488	1731.0	17.4
Main course	Hauptgericht	Hauptgericht	35	0.8	0.8
Appetizer	Vorspeise	Vorspeise	72	1.6	1.5
Dessert	Nachtisch	Dessert	193	4.2	6.5
Pastry	Backwaren	Gebäck	165	3.6	0.3

buttons, the best matching concept from the DBpedia ontology was identified and the number of links to it is used as measure of relevance of the concept (similar to PageRank; Page et al. 1999). As can be seen in Table 7.1, the buttons in the nationality group are two to three magnitudes more relevant than the buttons in the type-of-dish group. The empirical analysis therefore unfolds around the differences between those two groups. The corresponding factor is called *concept relevance* in the following, nationality corresponds to high concept relevance and type-of-dish to low relevance.

In order to create a reference for comparison to the analyses from the previous chapter, *goal relevance* (see Sect. 6.1) was introduced as second factor for the error analyses. This was operationalized as follows: The recipe search tasks featured trials where the participants were asked to change from one to another selection within the same group, e.g., "change the nationality from German to French and select Ratatouille". In case of the complex UI (see Fig. 6.4), this meant that the user had to perform two clicks, one to uncheck German and one to check French.[4] Of these two clicks, only the second one is directly connected to the overall goal (here: selecting Ratatouille). This yields a binary goal relevance factor with checking a search attribute being high goal relevance and unchecking a previously selected attribute being low goal relevance.

7.1.2   Results

Across the three experiments, there was a total of 4583 clicks on the eight search attribute buttons under investigation. 113 (2.5%) thereof were omissions and 130 (2.8%) were intrusions. The results for the ontology-based predictions are as follows.

---

[4] As the simple UI does not show the attributes and the search results side-by-side, this situation did not appear there.

Table 7.2 Linear mixed model results for the click time analysis (Experiments 2–4)

Factor name	Estimate	t	df	p
Fitts' index of difficulty in bit	209 $\frac{ms}{bit}$	10.10	58.4	<.001[a]
Device (tablet vs. screen)	114 ms	3.36	44.3	.002[a]
Attr. group (Dish vs. nationality)	83 ms	3.05	1801.2	.002[a]

Note: **highly significant (p < .01); *significant (p < .05); [†]marginally significant (p < .10)

Table 7.3 Mixed logit model results for errors depending on concept relevance, i.e., nationality (high) vs. other options (low). Experiment 2–4. OR: Odds ratio

Error class factor	OR	95% CI of OR		z	p
Omissions					
Low concept relevance	2.16	1.44	3.34	3.65	<.001[a]
Low goal relevance (uncheck task)	4.22	2.69	6.48	6.51	<.001[a]
Intrusions					
Low concept relevance	1.98	1.34	2.99	3.46	<.001[a]
Low goal relevance (uncheck task)	0.13	0.02	0.41	−2.95	.003[a]
All errors					
Low concept relevance	2.16	1.61	2.91	5.20	<.001[a]
Low goal relevance (uncheck task)	1.60	1.05	2.38	2.32	.021[b]

Note: **highly significant (p < .01); *significant (p < .05); [†]marginally significant (p < .10)

Task Completion Time

All clicks with substantial wait time (due to task instruction or system response) were excluded from the analysis. The remaining 2326 clicks still differ in the necessary accuracy of the finger movement which is strongly related to the time needed to perform the movement as formulated in Fitts' law (Fitts 1954). Individual differences in motor performance were large, and the device used (tablet vs. large screen) also had an effect on the click time. The subjects were therefore added as random factor to the analysis (with device and Fitts-slope within subject). The click time was analyzed using a linear mixed model (Bates et al. 2013), fixed effects were tested for significance using the Satterthwaite approximation for degrees of freedom. Results are given in Table 7.2. Besides the expected effects of Fitts' law and device, there was a significant difference between the buttons for concept relevance, with the low relevance group (i.e., type-of-dish) needing approximately 80 ms longer.

Omissions and Intrusions

If those 80 ms are caused by lack of activation (as predicted by the MFG), then this lack of activation should cause more omissions for the type-of-dish group (low relevance) and more intrusions for the nationality group (high relevance). Goal relevance as operationalized based on the uncheck task was added to the design to allow assessments of the relative strength of both effects. The analysis was performed using mixed logit models (subject within experiment as random effects). The results are shown in Table 7.3.

7.1.3   Results Discussion

Compared to the previous analysis in Halbrügge et al. (2015a), most effects are confirmed. Only the effect of concept relevance on intrusions points into the opposite direction, but this does not necessarily contradict the theory. The MFG explains intrusions by interference with earlier subgoals that are still present in memory. In the context of the experiment presented here, those intruding subgoals are memory clutter from already completed trials (see also the discussion in Sect. 6.9). In experimental design terms, this is called a carry-over effect (Westermann 2000). Due to the order of trials being randomized between subjects, intrusions should not happen on a general, but on a subject-specific level.

Descriptively, the effect of concept relevance is smaller than the goal relevance effect, but the confidence intervals still overlap. Compared to the general effects of device-orientation and task necessity presented in Table 6.6 on page 83, the impact of concept relevance is comparable to the main effect of task necessity.

7.1.4   Cognitive Model

The empirical results above back the theoretical assumptions of the cognitive model presented in Sect. 6.6 qualitatively, i.e., they are in line with the idea of goal chunks in declarative memory that are subject to semantic priming. Can this also be proved *quantitatively* as well? The following analysis explores how the predicted behavior of the (knowledge-free) model from Sect. 6.6 changes through the addition of concept priming. This is achieved by introducing pieces of information from Wikipedia to ACT-R's declarative memory system (Salvucci 2014) following the basic assumption that the short-lived subgoal chunks that are created by the model are subject to semantic priming from long-living general concepts.

Implementation

How much priming can be expected, based on the information that is available within DBpedia? Here, the inlink count is used as measure of the relevance of a concept. In ACT-R, this needs to be translated into an activation value of the chunk that represents the concept (i.e., Wikipedia article). Temporal decay of activation is modeled in ACT-R using the power law of forgetting (Anderson et al. 2004). Salvucci (2014) has applied this law to the concepts within DBpedia, assuming that they have been created long ago and the number of inlinks represents the number of presentations of the corresponding chunk. The base activation B can be determined from inlink count n as follows[5]

$$B = \ln(2n) \tag{7.1}$$

---

[5]Side remark: This resembles the log (word frequency) measure that is often used in computer linguistics (Brysbaert et al. 2011).

While Salvucci's rationale is generally reasonable, deriving the activation from the raw inlink count may be a little too straightforward. Numerically, it creates very high activation values. And as the total number of entries varies between the language variations of DBpedia, switching language (or ontology) would mean changing the general activation level.[6] In the special case presented here, the use of (7.1) caused erratic behavior because the high amount of ontology-based activation overrode all other activation processes (i.e., activation noise and mismatch penalties for partial matching of chunks). This was solved by introducing a small factor c that scales the inlink count down to usable values. Together with ACT-R's minimum activation constant blc, this results in the following equation

$$B = \max(\ln(c \cdot n), blc) \tag{7.2}$$

How is the semantic priming to subgoal chunks finally achieved? The declarative memory module of ACT-R 6 only allows priming from buffers ('working memory') to declarative ('long term') memory. Because of this, a 'hook' function had to be introduced that modifies the activation of every subgoal chunk whenever it enters long term memory according to the general concept that is related to the goal chunk.[7]

Goodness of Fit
The model was run 300 times with concept priming disabled ($c = 0$), and 300 times with priming enabled ($c = .005$, resulting average base activation of the eight concepts $M_B = 2.4$, $SD_B = 3.4$). For both conditions, task completion time and error predictions were computed for each button and were compared to the empirical observations. The effect of the needed accuracy of the finger move was eliminated based on Fitts' law, using a linear mixed model with subject as random factor (Bates et al. 2013) for the empirical data and a linear regression for the model data, as the Fitts' parameters were not varied during the simulation runs. Correlations between the respective residuals are given in Table 7.4. Omission and intrusion rates per button were correlated without further preprocessing.

The results are given alongside R^2 and RMSE in Table 7.4. While the goodness-of-fit with R^2 constantly below .4 and substantial RMSE is not overwhelming, the difference to the baseline is worth discussion. The model without concept priming displays medium to strong negative correlations between its predictions and the empirical values, meaning that the baseline model is even worse than chance. The corresponding regression lines are displayed on the upper part of Fig. 7.2. When concept priming is added, all three dependent variables show substantial positive correlations between observed and predicted values. The difference between the correlations is very large, i.e., always above .75.

---

[6]Example: The English DBpedia (version 3.9) is 2.5 to 3 times larger than the German one. "Intelligence" has 1022 inlinks in the English DBpedia, but "Intelligenz" has only 445 inlinks in the German one.

[7]The source code is available on GitHub (doi: 10.5281/zenodo.53200).

Table 7.4 Correlations between the empirical data and the model predictions (Experiment 2–4)

Dependent variable	$r_{baseline}$	$r_{priming}$	Δr	$R^2_{priming}$	$RMSE_{prim.}$
Execution time (residual)	−0.302	.615	.773	.378	81 ms
Omission rate	−0.493	.500	.797	.250	.019
Intrusion rate	−0.797	.275	.879	.076	.017

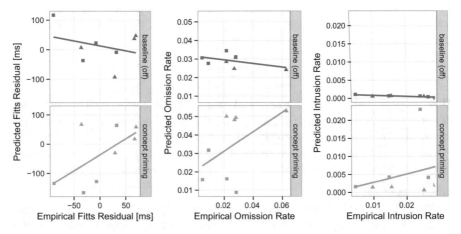

Fig. 7.2 Click time residuals after Fitts' law regression, intrusion and omission rates of the cognitive model with and without priming from the DBpedia concepts (experiments 2–4). Negative slopes of the regression line mean worse than chance predictions. Positive slopes mean better than chance predictions. Squares denote buttons of group "nationality" (high concept relevance), triangles denote "type of dish" (low concept relevance)

The positive correlation for intrusions is especially noteworthy as there was an opposite effect in the empirical analysis above (Table 7.3). If the hypothesis of intrusions being caused by leftovers from previous trials with additional priming from ontology-based concepts holds, then this result underlines the benefits of adding ontologies to cognitive architectures. A closer look at Fig. 7.2 reveals that the correlation for intrusions is highly dependent of a single outlier, the results should therefore be interpreted with care. It nevertheless confirms a similar result from Halbrügge et al. (2015a).

7.1.5   Discussion

Adding concepts from DBpedia (Lehmann et al. 2015) to the declarative knowledge of the cognitive model and modulating the activation of these concepts based on the number of links inside DBpedia that point to them allowed not only to reproduce the time and omission rate differences, but to some extent also the rates of intrusions. While the prediction of execution time and omissions mainly lies within the ontology, intrusions can only be explained by the combination of cognitive model and ontology, highlighting the synergy between both.

It is also informative to compare the approach to research on information foraging, namely SNIF-ACT (Fu and Pirolli 2007, see also the discussion of CogTool Explorer in Sect. 4.3.1). This system uses activation values that are estimated from word frequencies in online text corpora, which would lead to general hypotheses similar to the ones given above. But beyond this, a closer look unveils interesting differences to the DBpedia approach. While word frequency and inlink count are highly correlated ($r = .73$ in the current scenario, see Table 7.1), the word frequency operationalization yields much smaller differences between the nationality vs. type-of-dish groups. Frequency based-approaches also need to remove highly frequent, but otherwise irrelevant words beforehand (e.g., "the", "and"). In Wikipedia, this relevance filter is already built into the system and no such kind of preprocessing is necessary. Empirically, the analysis yielded inconclusive results when using word frequency in a large subtitle corpus (Brysbaert et al. 2011) instead of Wikipedia inlink count as concept activation estimate.

While the combination of cognitive model and ontology provides some stimulating results, it also has some downsides and limitations. First of all, the small number of observed errors leads to much uncertainty regarding the computed intrusion and omission rates. Especially in case of intrusions, the empirical basis is rather weak. The goodness-of-fit is highly dependent on outliers. While some of these matches the high-level predictions given in the introduction (e.g., "German" being more prone to intrusions), others point towards a conceptual weakness of the model (e.g., "Pastry" showing many intrusions in the empirical data although having just a few inlinks). The "Pastry" intrusions mainly happened during trials with the target recipes baked apples ("Bratäpfel") and baked bananas ("Gebackene Bananen"). One could speculate that those recipes have primed the type-of-dish attribute that is linked to baking. This kind of semantic priming is currently not covered by the system, as the ontology-backed activation is only applied statically. Allowing dynamic activation flow between the subgoals of the current action sequence is technically impossible in the current version of ACT-R. The following section will present a different approach to overcome this limitation.

With regards to automated usability evaluation, the results reported in this section—while being preliminary—point towards possible ways to evaluate the appropriateness of the concepts used in an interface. This is currently only covered by usability evaluation methods that involve actual users (e.g., CASSM; Blandford et al. 2008).

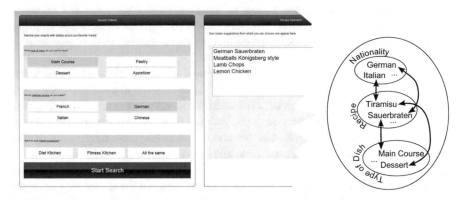

Fig. 7.3 English version of the kitchen assistant on the left; representation within LTMC on the right. Circles and text denote objects within their categories, arrows denote relationships. Activation spreads along the arrows. Originally published in Halbrügge and Schultheis (2016)

7.2   Modeling Application Knowledge with LTMC

In order to overcome the limitations of the ACT-R approach presented in the previous section, a reduced version of the error model has been implemented in a new system that uses an alternative model of human long-term memory, LTMC, for knowledge representation.

7.2.1   LTMC

LTMC (Schultheis et al. 2006, 2007) represents knowledge about the world as a semantic network of nodes. These nodes consist of their semantic content and their numerical activation. Links between nodes represent associations (see Fig. 7.3, right). The activation of a node stems from three influences: base level activation, noise, and spreading activation. In the current context, spreading activation is of highest importance. It distributes activation from goal nodes (e.g., "Sauerbraten") to related concepts within LTMC (here: "German" and "Main Course").

7.2.2   Method

The implications of semantic priming on human error were explored based on data that had been collected during Experiment 2 (Sect. 6.5). For the current analysis, the semantic mappings between recipes and search attributes were transformed to an LTMC object-relation-graph (Fig. 7.3, right). MeMo (Engelbrecht et al. 2008, see

Sect. 4.3.4) was used to represent the action sequences of the experimental procedure as chains of subgoals (e.g., "German", "Sauerbraten"). Each subgoal was augmented with a numerical activation value. LTMC was then used to modulate these activation values through activation spreading, i.e., in terms of semantic priming, the retrieval of the subgoal "Sauerbraten" was treated similarly to the retrieval of the generic "Sauerbraten" chunk from long-term memory. If the resulting activation was below a fixed threshold, the retrieval of the subgoal failed and the corresponding action was not carried out.

7.2.3 Results

The model's omission rates for different elements of the UI are given in Fig. 7.4 (100 model runs); the goodness-of-fit is satisfactory with $R^2 = .73$ and RMSE $= .0073$; the MLSD is 4.0.

Especially noteworthy are the low model omission rates for 'toggle attribute' and 'select recipe'. Without semantic priming, these predictions would be on the level of selecting and checking off ingredients. The low omission rates for 'start search' and 'next screen' on the other hand are due to visual priming during cue-seeking.

7.2.4 Discussion

The current section presented a model of sequential action in a household environment that was different from the ACT-R-based models from the previous chapters. By modulating the activation of individual subgoals based on semantic priming within LTMC, a good fit to previously recorded data was achieved. In principle, these aver-

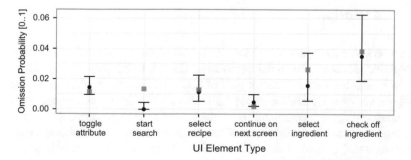

Fig. 7.4 Omission probabilities for different UI elements. ■ denote model predictions, error bars are 95% confidence intervals using the Agresti-Coull method

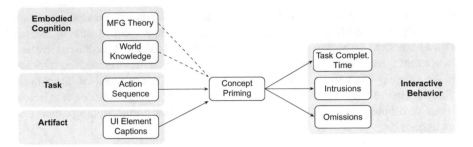

Fig. 7.5 Interactive behavior as a function of concept relevance

aged predictions could have been produced using the Wikipedia-based approach presented in Sect. 7.1 as well, but using LTMC has several advantages. First, LTMC's assumptions seem to better fit the mechanisms underlying human semantic priming, because ACT-R's activation mechanism had to be overridden (using a 'hook' function) to achieve similar effects in Sect. 7.1.4. Second, the priming in the previous approach was static, i.e., a subgoal 'German' received the same amount of additional priming regardless of the current task context. As the experimental procedure also contained tasks like "Switch from German to Italian and select Tiramisu", this approach led to the improbable result of 'German' being strongly primed during the 'Tiramisu' trial. This does not happen using the combined MeMo-LTMC solution (see arrows in Fig. 7.3).

The validity of the current approach is nevertheless limited. Because LTMC is only containing knowledge extracted from the application (instead of an external ontology), the priming results sometimes do not match the participants' conceptions. An example is the 'Ratatouille' recipe which is flagged as 'main dish' in the kitchen assistant. Several participants strongly objected this view during the experiment.

7.3   Conclusion

The current chapter explored how knowledge about the world can be integrated with a cognitive model of sequential behavior. Although being preliminary, the results can provide guidance to the larger goal of this work, i.e., the elicitation of usability problems during early development stages. First, while there was a significant effect of concept relevance (operationalized as number of links to the concept in Wikipedia, Lehmann et al. 2015), the influence of relevance on errors and task completion time was comparatively small (OR $\approx$ 2, see Table 7.3). This would probably improve with a better fitting ontology, e.g., one of cooking recipes.

Nevertheless, the results establish another link function between the ETA triad and interactive behavior as visualized in Fig. 7.5.

Second, intrusions seem to depend more on the sequence of the tasks than on the computer system (see also Sect. 2.3.2). This calls for focusing on omissions for AUE systems.

Finally, long-standing cognitive architectures like ACT-R (Anderson 2007, see Sect. 4.2) struggle with the demands of ontology integration. Building a dedicated system in a current programming language yielded faster and better results.

This closes the empirical part of this work. The next chapter presents the actual application of the empirical and modeling results from the previous three chapters in the development of an actual AUE system.

Part III
Application and Evaluation

Chapter 8
A Deeply Integrated System for Introspection-Based Error Prediction

In this chapter[1]:

- Combine the introspection-based predictions from Chap. 6 and the ontology-based approach from Chap. 7 into an integrated system for the automated usability evaluation of model-based applications.
- Evaluation of the system using new data. Does it generalize well?

After the theoretical basis has been layed out and the derived cognitive model has been sufficiently validated, the focus shall now shift to the question of *applicability* of the approach. Are goal relevance, task necessity and formalized world knowledge only instrumental for understanding human action control, or can these concepts also help designing better systems? More specific: Are predictions based on the cognitive model *automatable*, and do they yield valid information that is *relevant* in the design process?

Based on theoretical considerations and the empirical evidence (see overview in Table 6.6), the goal of the AUE system is the prediction of *omission* errors, only. This is done for the following reasons:

- On the theoretic level (Gray 2000; Altmann and Trafton 2002, see Sect. 2.3), only omissions are a direct function of the UI design (e.g., through lack of goal relevance). In theory, intrusions are caused by the interaction of design and task sequence (i.e., by 'memory clutter' from previous tasks being primed by UI elements). This view has be reinforced by the results of the data reanalysis in Sect. 7.1.
- In Experiment 2 and 3—which featured the highest ecological validity—the base rate for omission errors was much higher than the intrusion rate.

---

[1]Parts of this chapter have already been published in Halbrügge et al. (2016).

© Springer International Publishing AG 2018
M. Halbrügge, *Predicting User Performance and Errors*, T-Labs Series
in Telecommunication Services, DOI 10.1007/978-3-319-60369-8_8

- Overall, the results for intrusions were inconclusive across the three error experiments. The effects for omissions on the other hand represent effects for all sequential errors well (see Table 6.6).

For reasons of completeness, intrusions are nevertheless included in the empirical analyses in this chapter.

8.1   Inferring Task Necessity and Goal Relevance From UI Meta-Information

A prerequisite of automatibility is that the input variables of the cognitive model must be derivable from the application under evaluation computationally (i.e., without intervention of an analyst). Usually, this is not the case for properties like device-orientation. While a human analyst can spot device-oriented UI elements easily, the same is very hard computationally, based only on the textual caption or visual look of a UI element.[2] This situation is different for runtime architectures of model-based applications like the MASP (see Sect. 3.2). By following the mutual links between the final UI and the underlying development models, information about the application becomes available that goes far beyond what is visible to the user.

With respect to goal relevance, device-oriented elements can be identified based on information on the AUI and the task model (CTT) layer. As device-orientation is characterized by not transferring any information from the user to the system, the AUI type of such an element must not be 'input' or 'choice', but 'command'. Furthermore, device-oriented elements are necessary to operate the device. This is reflected on the task model level (highest level of abstraction in the CAMELEON framework) by the corresponding 'interaction task' nodes having an enabling relation to 'application task' nodes, i.e., the selection of the AUI command by the user leads to the execution of some internal operation of the application. This rationale is illustrated in Fig. 6.4 on page 62. Note that the task analysis techniques commonly applied by usability experts (e.g., Hierarchical Task Analysis; Kirwan and Ainsworth 1992) do not yield sufficiently rich information as they do not incorporate user and application tasks (which is a specific property of CTT-style models; Paternò 2003).

Task necessity on the other hand can not be deduced directly from the abstract models as it often emerges on the FUI stage, only (e.g., when more UI screens are added to adapt to a device with a smaller form factor; see Fig. 6.4). This is also the reason why simulation using the runtime system is necessary. While device orientation can be derived statically, task necessity is only implicitly represented in the computational code of the FUI. Both factors need to be taken together to achieve good predictions (see discussion in Sect. 6.9).

---

[2]Current machine learning techniques might be able to overcome this, but this approach has not been tried, yet.

8.2 Integrated System

While the ACT-R approach is helpful to advance psychological theory, it does not scale well to applied scenarios (see discussion in Sect. 7.3). Therefore, the core components of the ACT-R model were replicated within the MeMo system that had already been integrated with the MASP framework that has been used for the experiments in the previous chapters.

In Sect. 4.4, an integrated system for automated usability evaluation of model-based applications was presented that accesses required information from relevant UI development models. This system is based on the MeMo workbench for usability testing (Engelbrecht et al. 2008, see Sect. 4.3.4). MeMo's user simulation progresses towards a given goal state by utilizing a combination of breadth-first and depth-first search on the state graph of the application under evaluation. By integrating MeMo with the Multi Access Service Platform (MASP; Blumendorf et al. 2010), a CAMELEON-conformant runtime framework, this state graph can be extracted directly from the application (see Fig. 8.1), thereby saving the analyst from creating a mock-up UI with MeMo (Quade 2015).

The integrated system for error prediction that has been created for error prediction extends on the system by Quade (2015) in several respects:

- In the old system, only the current state of the application was available to MeMo and no path search was performed (see Fig. 4.4). This was sufficient because the evaluation goal were task completion times and the action sequence performed by the user model was fixed.

  In the new version, potential user errors lead to some randomness in the interactions that are chosen by the user model. In order to handle this computationally, a (partial) state transition graph of the application had to be derived from the MASP models and MeMo's path search is used to determine an action sequence that leads to a specified goal state (see Fig. 8.1).

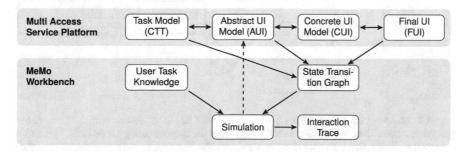

Fig. 8.1 Structure of the integrated MASP-MeMo system for error prediction. Information flow is denoted by solid arrows. The MeMo simulation engine applies interactions on the AUI level of the MASP (*dashed arrow*). Originally published in Halbrügge et al. (2016). © 2016 Association for Computing Machinery, Inc. Reprinted by permission

- As stated above, information on the CTT level is necessary to determine device-orientation. For this reason, a new link from the task model within the MASP to the MeMo simulation was established.
- In order to make the integrated system capable of errors, its until then optimally behaving (path search based) user simulation had to be modified in a reasonable manner. This was done based on the MFG theory as introduced in Sect. 2.3.3 and on the results from the experiments reported in the previous chapters. While path search on the state graph of the application is still used to find an optimal sequence of interactions to attain a goal, the individual interactions (i.e., arrows in Figs. 8.2 and 8.3) in the resulting sequence are now subject to memory activation.

The actual simulation is driven by the tasks of the users. A single task is defined by its start state (e.g., the home screen of the UI), its goal state, and an arbitrary number of *user task knowledge* items. The task knowledge is the set of information items that the users have to transmit to the software application in order to attain their goals. The individual items in the user task knowledge are applied on the path from start state to goal state by comparing them to the captions of the currently visible elements of the UI. If an element matches, a corresponding interaction is performed by the simulation (dashed arrow in Fig. 8.1). Such interactions usually correspond to task-oriented user goals.

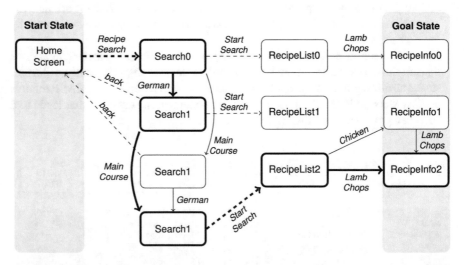

Fig. 8.2 MeMo system model (detail) of a recipe search. Nodes denote states, arrows denote transitions. Transitions are labeled with the caption of the corresponding UI element. Thick lines represent an interaction path simulated by MeMo for the user task knowledge {"German", "Main Dish", "Lamb Chops"}. Note that MeMo has automatically added the interaction steps "Recipe Search" and "Start Search" through path search. Semantic processing within MeMo has mapped the user task knowledge "Main Dish" to the UI element "Main Course" (Hanisch 2014). Originally published in Halbrügge et al. (2016). © 2016 Association for Computing Machinery, Inc. Reprinted by permission

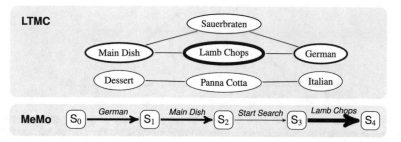

Fig. 8.3 Knowledge representation and application of spreading activation within LTMC and application to MeMo simulation. Activation is visualized as node border thickness in LTMC and as arrow thickness in MeMo. By highly activation the overall goal of the sequence ('Lamb Chops'), semantically connected nodes ('German', 'Main Dish') receive priming. The resulting activation values are applied during MeMo's interaction simulation. Note that the device-oriented subgoal 'Start Search' does not receive any priming at all. Originally published in Halbrügge et al. (2016). © 2016 Association for Computing Machinery, Inc. Reprinted by permission

Additional steps may be necessary to proceed towards the goal state, e.g., navigation to subsequent UI screens. These are found by the MeMo based on path search on a state transition graph that it derives from the task model, AUI model, and FUI of the MASP application (see Fig. 8.1). Interactions without corresponding user task knowledge items correspond to device-oriented user goals.

An example of how the system works is visualized in Fig. 8.2. It is based on the task "Search for German main dishes and select lamb chops" (see UI in Fig. 3.2). The verbal task description is first divided into three items: 'German', 'main dish', and 'lamb chops'. The start state is the home screen of the UI, the goal state is a screen that gives recipe information like preparation time and calorie content.

8.2.1 Computation of Subgoal Activation

Each element of the task description that the model is following (i.e., each subgoal) was extended by a numerical activation value. For reasons of simplicity, these activation values are not computed using ACT-R's sophisticated activation formulae, but are drawn from a standard Gaussian random variable instead (see Fig. 8.4). Omissions occur when the activation of a goal (i.e., element of the task description) happens to fall below a fixed retrieval threshold rt. Visual priming while searching the screen when using the knowledge-in-the-world strategy is achieved by adding a fixed visual priming constant vp.

Task Priming is modeled as a positive numerical value that is added to the goal's activation, thereby reducing the probability of retrieval failure. It represents activation spreading from the overall goal (e.g., looking up the 'lamb chops' recipe) to a subgoal leading to it (e.g., using a 'main dish' search attribute). The LTMC-approach from

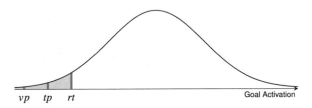

Fig. 8.4 Mapping of empirical omission rates to the parameters of the cognitive user model. Activation is assumed to be a normally distributed random variable. The area under the curve left of one of the lines indicates the omission probability of a corresponding goal. Originally published in Halbrügge et al. (2016). © 2016 Association for Computing Machinery, Inc. Reprinted by permission

Sect. 7.2 is used to compute the amount of task priming. LMTC's main purpose is to represent knowledge about the world as semantic networks (Schultheis et al. 2006). The integrated system as presented here uses it to represent the user knowledge about the current system, only. This was achieved by adding all recipes contained within the kitchen assistant to LTMC together with their attribute mapping, i.e., the semantic network was built up from facts like 'Panna Cotta is a dessert' or 'Sauerbraten is a main dish'.

Within the network, a specific dish (e.g., 'Lamb Chops') spreads activation to its attributes ('German' and 'main dish') but not to the other nationalities or types of dish. Search attributes accordingly spread activation to recipes, but as more recipes belong to a single search attribute than attributes to recipes, the amount of spreading is smaller in this direction. For each action of the user model, the LTMC module is run to compute the amount of priming that the goal that corresponds to the step receives. The result of the computation within LTMC is applied after rescaling it by the maximum task priming constant tp. The process is visualized in Fig. 8.3.

8.2.2 Parameter Fitting Procedure

Because of their higher ecological validity compared to Experiment 3, only the combined results of both Experiment 1 and 2 serve as empirical basis of the user model. The model has three free parameters, the value of each can be directly estimated from data.

- The retrieval threshold rt below which a task is forgotten is estimated from the data using the omission probability of device-oriented non-obligatory tasks ($p = .037, z = -1.78$, see Fig. 8.4). This task category does neither receive task priming, nor is it enforced by the application logic. As the users must completely rely on their memory in this case, the empirical omission rate should be a good estimator of the retrieval threshold.

- The amount of task priming tp is estimated as the difference in omission rates between task-oriented and device-oriented non-obligatory steps ($\Delta z = 0.48$).
- Visual priming vp is estimated using the difference between obligatory and non-obligatory device-oriented tasks ($\Delta z = 1.02$).

The fit of the resulting model was assessed by performing 100 model runs with the task set used in Experiment 1. The fit to the training data is very good, $R^2 = .99$, RMSE $= .0038$. The Maximum Likely Scaled Difference (MLSD; Stewart and West 2010), that takes the variability of the data into account, is 1.94. Being so close to its theoretical minimum of 1 means that the model should not be refined without the danger of overfitting (see Sect. 5.2).

8.3 Validation Study (Experiment 5)

In order to test the validity of the integrated system and the generalizability of the user model, a new experiment was designed using a different application from a similar domain and new types of user tasks. The new application is a MASP-based health assistant that has been developed as part of a health information system for migrants (Plumbaum et al. 2014, see Fig. 8.5).

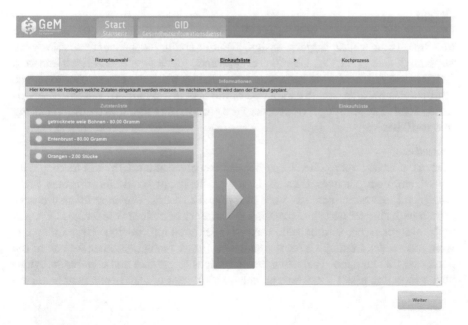

Fig. 8.5 Screenshot of the German version of the health assistant

The health assistant contains a recipe search similar to the kitchen assistant used in the previous experiments. In contrast to the kitchen assistant, the health assistant's UI is built around the health conditions of its users. It also features a finer grained shopping list generator that can handle several personalized lists at a time.

8.3.1 Method

Participants
The experiment was conducted in July and August 2015. 30 participants, 15 men and 15 women, aged between 19 and 56 (M = 33.7, SD = 8.7), were recruited from the paid participant pool of Technische Universität Berlin. Written consent was obtained from all participants.

Materials
The health assistant was displayed on a 23" (58.4 cm) monitor with optical sensor 'touch' technology similar to the one used during Experiments 3 and 4 and on the 10" (25.7 cm) tablet used before. All devices operated in landscape mode. User actions were again recorded by the computer system and additionally videotaped for error classification.

Design
Dropping the comparison of different UI versions, the experiment could be reduced to a three-factor within-subjects design. The remaining independent variables were physical device (screen vs. tablet), goal relevance (device- vs. task-orientation) and task necessity. A new set of tasks was generated by a different researcher to reduce implicit bias towards simple paraphrase of the task instructions used during Experiments 2 to 4. The resulting 35 user tasks were grouped into four trial blocks consisting of 8 to 10 trials.

Procedure
After a quick warm-up game, the participants received about 10 min of training with the system. Four personas, three of them with the health conditions diabetes, pregnancy, and lactose intolerance, were introduced to them. The experimental blocks started with simpler tasks like counting the number of lactose-free recipes available in the system, or comparing health-related nutritional information. The participants were then walked through a background story (e.g., planning dinner for one of the personas) that comprised selecting which recipes to prepare and creating individualized shopping lists.[3] The whole procedure took approximately one hour.

---

[3]The full instructions are available at Zenodo (doi: 10.5281/zenodo.268596).

Table 8.1   Mixed logit model results for Experiment 5. OR: Odds ratio for errors

Error class	Factor name	OR	95% CI of OR		z	p
Omissions						
	Tablet vs. Screen	1.04	0.83	1.32	0.36	.718
	Device-oriented step	1.68	0.77	4.22	1.23	.219
	Non-obligatory step	2.89	1.46	6.83	2.76	.006**
	Interaction Dev.-Or. × Non-obligatory	0.59	0.22	1.37	−1.17	.243
Intrusions						
	Tablet vs. Screen	1.50	1.01	2.26	2.01	.044*
	Device-oriented step	1.68	0.25	32.90	0.47	.638
	Non-obligatory step	8.86	1.96	156.71	2.21	.027*
	Interaction Dev.-Or. × Non-obligatory	0.12	0.01	1.03	−1.75	.079†
All errors						
	Tablet vs. Screen	1.15	0.94	1.41	1.33	.185
	Device-oriented step	1.69	0.81	3.97	1.32	.188
	Non-obligatory step	3.73	1.96	8.27	3.66	<.001**
	Interaction Dev.-Or. × Non-obligatory	0.44	0.18	0.96	−1.96	.050†

Note **highly significant ($p < .01$); *significant ($p < .05$); †marginally significant ($p < .10$)

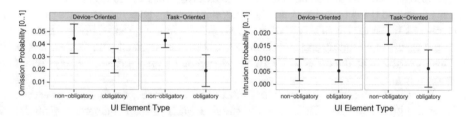

Fig. 8.6   Error probabilities for Experiment 5. Error bars are 95% confidence intervals using the Agresti-Coull method

8.3.2   Results

There were 7663 clicks in total, 103 (1.3%) thereof were intrusions and 297 (3.9%) were omissions. On the first try, 25.9% of the trials were not completed correctly by the participants. Because there was an increased intrusion rate on the tablet, the device factor is included in the analysis. The GLMM results with subject as random factor are shown in Table 8.1, means and confidence intervals for omissions and intrusions are shown in Fig. 8.6.

8.3.3   Results Discussion

The overall error rate is higher than in most of the previous experiments, but still within the usual range of 5% for procedural error (Reason 1990). The biggest difference is the high omission rate for non-obligatory task-oriented goals that is on the same level as the omission rate of their device-oriented counterpart. This result is rather unexpected because the previous experiments produced the opposite pattern in line with the task-priming assumption (see discussion in Sect. 6.9).

The user tasks connected to this category are a) selecting ingredients and b) toggling search attributes. The high omission rate is most probably caused by the use of rather uncommon ingredients (e.g., soy-drink, quince purée) and search attributes (e.g., lactose intolerance).

8.4   Model Fit

In order to assess the generality of the model, all model parameters were kept at the values that had been estimated from previous data (see above and Fig. 8.4). In order to ensure a conservative estimation of the fit, clicks connected to manipulating the number of servings were excluded from the analysis. These fall into the non-obligatory task-oriented category and showed the smallest error rates in the experiment (0.7% omissions and 0.4% intrusions). As the integrated system is technically incapable of erroneous behavior for this kind of click, keeping it in the goodness-of-fit analysis would result in a slightly inflated fit measure. From an applied perspective, this is a lesser problem as the order of UI elements ranked by error proneness stays intact (see Table 8.2).

The model completed the health assistant trials 100 times, the fit to the new data is good with $R^2 = .70$. While the rather high RMSE of .021 indicates the model missing the generally higher omission rate in the validation experiment, the MLSD value of 2.6 indicates a reasonable fit given the uncertainty in the empirical data (see Fig. 8.7). Despite not anticipating the overall increased omission rate, one can see that the higher omission rate of task-oriented non-obligatory tasks is well captured by the model.

More important than the quantitative fit is the qualitative usefulness of the model: Can it facilitate the UI design process, i.e., does the model identify UI elements that are especially prone to errors? This question was approached by comparing the omission rates for different types of UI elements on a rank basis. As presented in Table 8.2, most ranks are matched between data and model. The biggest difference occurs for *search attribute* buttons, which were empirically the 2nd most forgotten type while the model predicted them being ranked 5th.

Especially noteworthy is that both model and empirical data show the highest omission rates for the *select ingredients* type of element. An exemplary screenshot

Table 8.2 Ranks of the empirical and predicted omission rates

UI element type	Device-oriented	Rank (data)	Rank (model)
Select ingredient	–	1st	1st
Toggle search attribute	–	2nd	5th
Move ingredient to shopping list	yes	3rd	2nd
Select persona	–	4th	4th
Assign ingredient to persona	yes	5th	3rd
Continue on next screen	yes	6th	6th
Select recipe	–	7th	7th
Increase number of servings	–	8th	

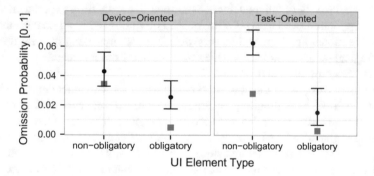

Fig. 8.7 Fit of the cognitive user model to the validation experiment. Error bars denote 95% confidence intervals of the empirical omission rates; ■ denote model predictions

of the health assistant's ingredients list is shown in Fig. 8.5. This element falls into the task-oriented category, i.e., it should be rather resilient to omissions following the task priming assumption. How does the model produce this prediction?

Three factors contribute to the high omission rate of ingredients: First, the ingredients appeared unusual to some of the participants and were only loosely connected to the recipes (e.g., tofu and teriyaki sauce for a pepper skewers recipe). The amount of task priming was therefore practically zero. Second, the trials concerning ingredients were rather long and often contained several ingredients. With every ingredient added to the list, the probability that (at least) one ingredient in a trial is forgotten becomes much higher. And finally, once a goal retrieval fails, the visual search for suitable elements as proposed in the *knowledge-in-the-world assumption* is a (partially) unsystematic process. In case of long ingredients lists, a distracting list entry can take over control and the forgotten goal never gets a chance to come into action.[4]

---

[4] Note that the first and the last factor are of general nature, only the co-occurrence of all three has led to the specific error pattern of the validation experiment.

8.5   Discussion

In the following, the integrated system is discussed from two perspectives: the validity of the user model and the benefits and limitations compared to existing approaches.

8.5.1   Validity of the Cognitive User Model

The generalizability of the model was fostered on several ways. First, the studies presented in the course of this work have been designed using real world applications and usage scenarios and have drawn participants mainly from non-student populations. The generality of the model was examined in a validation experiment using a different application and new set of tasks. Although this led to an unexpected error pattern, the model fits the new data very well. Second, by basing the user model on psychological theory, it builds upon the results of many other researchers instead just on the studies presented in this work.

The model extends on the activation-based Memory for Goals theory (Altmann and Trafton 2002) by highlighting the importance of external cues during sequential action. Internal cues are divided into task-oriented, i.e., steps that directly contribute to the users' goals, and device-oriented ones (Ament et al. 2013). Together with the assumption that only task-oriented steps receive additional priming from the user's overall goal, this allows not only to explain the data, but also provides an acceptable explanation of how device-orientation actually affects sequential behavior.

The original MFG model is based on the ACT-R theory (Anderson et al. 2004) which provides a psychologically plausible framework for the computation of activation values. In the integrated system, the cognitive plausibility provided by ACT-R was traded for the practical applicability of the model to design questions in HCI. By reducing a previously validated ACT-R model to a simple probabilistic model with few parameters, it was possible to integrate the model into the MASP system, a runtime architecture for model-based applications. Thereby, no mock-up of the application is necessary to perform the evaluation, and the automatic identification of device-oriented (i.e., potentially error-prone) task steps is possible through inspection of UI meta-information (i.e., AUI and CTT model) provided by the MASP.

The error model has several limitations. First, the model only covers expert behavior. The initial formation of the task sequence by novice human users is beyond its capabilities. The model also does not have any general knowledge and therefore cannot account for errors caused by the UI design violating general expectations of its users towards computer systems (e.g., about the functionality of a 'home' or a 'back' option). Second, the connection to a specific MBUID system limits the applicability of the model to applications that are developed within that system. Third, the domain knowledge of the user model within LTMC is currently restricted to the information that is used and provided by the application. Thereby, it is not possible to spot inconsistencies with user expectations.[5]

8.5.2 Comparison to Other Approaches

While the system presented here is the first to combine MBUID with psychologically plausible user models for error prediction, the prevention of user error based on UI meta-information has been proposed before.

Sect. 4.4 has given an overview of several methods that rely on inspection of the task model of a MBUID application. These differ from the approach presented here in that they lack the automation and integration that the MASP-MeMo-LTMC system provides. Also, the data show that the information on the level of the task model is not sufficient for the prediction of omission errors (see Sect. 8.1). Many device-oriented tasks, e.g., navigation to a subsequent page, are not even represented on the AUI model level but introduced later when the actual FUI is developed.

In recent years, formal verification has been proposed to ensure error tolerant systems (Bolton et al. 2012; Rukšėnas et al. 2014, see Sect. 4.3.3). Due to the formal nature of the approach, its application requires highly specific knowledge during the modeling of the system under evaluation. The software used for the verification task is computationally heavy and does not provide the automation that is possible by the integration with an MBUID system. The approach may nonetheless be worth the effort in safety-critical scenarios, e.g., as part of the general approval process in the medical domain. In contrast, the AUE approach presented here is meant to provide early predictions of usability issues during ongoing development processes that are applied regularly, e.g., as part of a continuous integration system. It is therefore designed to be easy to use and meant to provide first hints about which part of an UI should be redesigned.

Common to all previous approaches discussed above is that they can not account for the problem of plasticity (Coutaz and Calvary 2012) of adapted UIs as they are needed for multi-target systems. The sheer number of possible devices and form factors renders inspection-based or verification-based approaches impossible if all UI variations are meant to be evaluated. The automation provided by integrated systems like the one presented here solves this problem.

---

[5]Example: The kitchen assistant contains a recipe for Ratatouille (French vegetable stew) and files it under main dish. During the usability studies, two participants voiced objections because they regarded Ratatouille as being a side dish, only.

8.6   Conclusion

The current chapter presented the integrated system for error prediction during early stages of UI development that builds upon the results of several previous chapters. Technically, the system presented here is based on the MASP-MeMo-system that has been presented in Sect. 4.4 and validated in Chap. 5. The scientific foundation of the error prediction and its validation has been laid out in Chap. 6. How the knowledge of the simulated users can be formally described and computationally represented was explored in Chap. 7. The error prediction system uses the knowledge system (LTMC; Schultheis et al. 2006) that has been validated in that chapter, as well.

A validation experiment using a different MASP-based application yielded a reasonable fit of the system's prediction to the empirical omission rates although the error pattern differed considerably from the ones presented in the previous chapters. Qualitatively, the system was able to spot most UI elements that were error prone (see Table 8.2).

The relatively good fit to the unexpected error pattern highlights the benefits of using a psychologically grounded user model. Purely data-driven approaches, e.g., statistical models, would most probably not have been able to produce similar fits as the higher order interactions observed here (unusual items × long tasks × many visual distractors) were not well represented in the previous data that was used for fitting the model.

Chapter 9
The Unknown User: Does Optimizing for Errors and Time Lead to More Likable Systems?

In this chapter:

- Third aspect of usability has not been covered yet: *user satisfaction.*
- Is it related to the concepts introduced in the previous chapters?
- Can it be predicted as part of an AUE system?

A general premise of this work as presented until here was that "humans strive to be efficient," and therefore "the user's activity can be predicted to a great extent from the system design" (John and Kieras 1996a, see Chap. 2). The current chapter will broaden this view by shifting the focus to satisfaction. User satisfaction is not only the result of perception, but also determined by emotions and their cognitive appraisals on the one hand (Thüring and Mahlke 2007), and comparison to prior expectations on the other hand (Möller 2010). For this reason, more factors that drive human behavior than striving for efficiency have to be added to the equation.

The current chapter does not aim to answer all questions regarding the predictability of user satisfaction in general, its main purpose is to put the applicability and scope of the integrated system presented in the previous chapter into perspective. The leading question is therefore: Given a usability prediction system based on goal relevance and task necessity (which are automatically derived through UI introspection), to which extent would this system be able to predict user satisfaction ratings as well? In the following, this question is approached empirically with a dedicated online study.

© Springer International Publishing AG 2018
M. Halbrügge, *Predicting User Performance and Errors*, T-Labs Series
in Telecommunication Services, DOI 10.1007/978-3-319-60369-8_9

9.1   Device-Orientation and User Satisfaction (Experiment 6)

While satisfaction measures had been included in Experiments 0 and 2, their analysis did not yield conclusive results. These experiments featured repeated measures designs, i.e., the participants gave satisfaction ratings after successive experiences with variants of the same application. This led to strong interactions with previous ratings which is most likely due to the participants rather rating their overall experience over the course of the study than only the latest one (Halbrügge and Engelbrecht 2015). Anecdotally, there was nevertheless evidence that at least some participants preferred one UI variant because it needed fewer (device-oriented) clicks to attain the same goal.[1]

To shed more light on the impact of device-orientation on user satisfaction, but at the same time avoiding the pitfalls of repeated measures, a new between-subjects experiment was designed. Because this design choice introduces more interpersonal variation into the analysis and because the expected effects were small (i.e., below 5% explained variance, Hornbæk and Law 2007), a large sample was needed. As this would have been too costly if following the laboratory procedures of the MASP-based Experiments 0–5, an online study with a dedicated application was developed instead. This application does not use the model-based approach, but this is also not necessary in this case. In order to keep the task connected to the kitchen assistant and the health assistant that were used before, the experiment uses a tag-based picture search which remotely resembles the recipe search functionality of the MASP-based assistants.

9.1.1   Method

Participants
The study was online for about 72 h between the 12th and the 15th of July 2016. All members of the paid participant pool of TU Berlin were invited by e-mail to take part in the study. The landing page received 851 individual visits while it was online. 472 participants completed both the experiment and the questionnaire. There were 293 (62.1%) women and 179 (37.9%) men, their age ranged from 18 to 72 ($M_{age} = 30.5$, $SD_{age} = 10.8$). Instead of a fixed compensation, a total of 100€ was awarded to three randomly drawn participants.

Materials
The experiment unfolded around a tag-based picture search. A total of 24 search tags were grouped into three columns titled object (e.g., food, building, animal), nation (e.g., Germany, France, USA), and abstract concept (e.g. joy, business, night-time), respectively. The pictures that were presented as search results were preview pictures

---

[1]This was the 'complex' UI version of Experiment 2, see Sect. 6.5.

of the commercial service Shutterstock[2] which also provided the search facility used for the experiment. For each search, up to nine results were presented in a 3-by-3 matrix on an overview page (Fig. 9.1, top). Depending on the experimental condition, the participants could select a picture by dragging it to a dedicated selection area on the screen, or were forced to visit a picture-specific detail page, first (Fig. 9.1, bottom). The detail page featured a larger view of the picture alongside its textual description from the database. Here, a selection was made by either dragging the larger view to the selection area or using a button below the picture (Fig. 9.1, bottom).

In order to minimize the effect of (emotional) priming through the semantic content of the pictures, the search results were not presented based on highest relevance (according to the current search), but were a random sample of all pictures in the database that conform to the current search. This should make sure that each participant would get a different set of pictures for each individual search, thereby reducing the possibility of general priming effects from pictures to emotion and satisfaction ratings.

During the experimental stage, all user actions were logged using JavaScript code on the client side. As the response time of the external picture search was noticeable (median time $= 1.5$ s), the time between the initiation of a search and the return of its result was logged as well.

In a subsequent stage, two questionnaires were applied using the HTML-based software TheFragebogen (Guse 2016). These were AttrakDiff Mini (Hassenzahl and Monk 2010) and meCUE (Minge and Riedel 2013) which is based on the CUE (Components of User Experience) model of user experience (Thüring and Mahlke 2007). In case of the meCUE questionnaire, the generic notion of a "product" that is part of every item was replaced by "picture search" to keep the focus on the experiences during the preceding experimental stage.

Design

A single-factor between-subjects design was implemented by manipulating the proportion of device-oriented steps on three levels.

- In the *drag-from-overview* condition with the smallest proportion of device-oriented steps, the detail view was not accessible and the participants were instructed to drag pictures directly to the selection area.
- In the *drag-from-detail-view* condition, dragging the picture from the overview was not possible. After clicking on a picture, its detail view appeared. The participants were instructed to drag the picture from there.
- In the *apply-button* condition, no dragging was possible. The participants were instructed to navigate to the detail view and use an "apply" (German: "Übernehmen") button that was part of it.

Dependent measures were user satisfaction ratings and interaction sequences. Age, gender, operating system used and the web browser application used by the participants were entered as covariates.

---

[2]http://www.shutterstock.com.

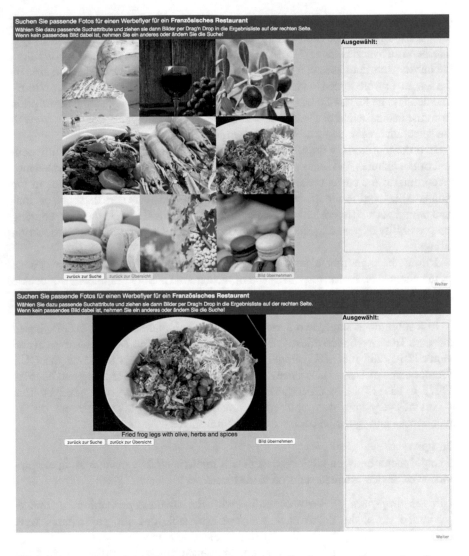

Fig. 9.1 Screenshots of the UI of the online experiment. *Top* Search results overview. *Bottom* Search result detail page (after clicking on one of the thumbnails of the overview). Images used under license from Shutterstock.com

Procedure

The experiment started with a greeting page that introduced the picture search and explained the money lottery that would happen afterwards.[3] The goal of the subsequent tasks was given as "imagine you were designing advertisement flyers and needed pictures for that". Then each participant went through the same three tasks, namely searching for pictures for flyers for a French restaurant, a public viewing

---

[3]The full instructions are available at Zenodo (doi:10.5281/zenodo.268596).

event of a soccer match, and the opening of a car repair shop. The AttrakDiff Mini and meCUE questionnaires were applied afterwards, followed by demographic questions and finally a text box for entering an e-mail that would take part in the lottery. On average, the total procedure took 15 min.

9.1.2 Results

10 participants (2.2%) were rejected because they did not fill out the questionnaire conscientiously, operationalized by the entropy generated by them being 3 standard deviations below the average (see also Naderi et al. 2015). 37 participants were excluded because they had completed the experiment on a mobile device (which was not suited for the procedure). Another 25 participants had to be excluded because of incomplete data, leaving a total of 400 cases (60% female, $M_{age} = 30.3$, $SD_{age} = 10.5$) for the analysis.

On average, the participants needed 9.3 min (SD = 6.0) to complete the experiment and performed 15 (SD = 20) individual picture searches during that time.

Questionnaires
The scales of both questionnaires were computed according to the respective manuals. Intercorrelations and Cronbach's alpha values are given in Table 9.1 for AttrakDiff and in Table 9.2 for the meCUE questionnaire. The reliabilities of the scales are overall satisfactory.

Compared to AttrakDiff, the meCUE questionnaire adds additional value because a) it uses a different question format (Likert scales instead of semantic differential), b) it adds the scales status (e.g., "the product gives me higher reputation") and commitment (e.g., "the product is like a friend to me") to the aspect of hedonic quality, c) intention of use (e.g., "if I could, I would use the product every day") and loyalty (e.g., "I would never exchange this product for another one") are added to measure the overall acceptance of the product, and d) emotions towards the product are covered as well (e.g., "the product makes me angry/happy"; all translations by the author).

Table 9.1 AttrakDiff Mini scale intercorrelations. Raw correlations below the diagonal, Cronbach's alpha in italics on the diagonal, correlations corrected for attenuation (Lord et al. 1968, Chap. 3) above the diagonal

		PQ	HQ	B	G
PQ	Pragmatic quality	*.790*	.518	.425	.689
HQ	Hedonic quality	.422	*.840*	.731	.815
B	Beauty	.378	.670	–	.573
G	Goodness	.613	.747	.573	–

Table 9.2 meCUE scale intercorrelations. Raw correlations below the diagonal, Cronbach's alpha in italics on the diagonal, correlations corrected for attenuation (Lord et al. 1968, Chap. 3) above the diagonal. For the four emotion scales, A+/− indicates high versus low arousal

		N	U	A	S	C	AP	DP	AN	DN	NI	L	O
N	Effectiveness	*.788*	.586	.800	.836	.585	.831	.722	−.708	−.597	.900	.802	.851
U	Efficiency	.479	*.848*	.385	.318	.130	.367	.353	−.579	−.580	.343	.285	.623
A	Aesthetics	.659	.328	*.859*	.696	.397	.740	.633	−.551	−.436	.704	.705	.731
S	Status	.652	.257	.567	*.771*	.812	.828	.725	−.417	−.206	.871	.842	.607
C	Commitment	.470	.109	.334	.646	*.820*	.672	.566	−.256	−.110	.702	.584	.360
AP	Pos. Emo. (A+)	.685	.314	.637	.676	.565	*.863*	.953	−.602	−.376	.926	.855	.710
DP	Pos. Emo. (A−)	.590	.299	.540	.586	.472	.815	*.847*	−.516	−.313	.810	.732	.632
AN	Neg. Emo. (A+)	−.595	−.505	−.483	−.347	−.220	−.529	−.449	*.896*	.907	−.563	−.483	−.736
DN	Neg. Emo. (A−)	−.449	−.453	−.343	−.154	−.084	−.296	−.244	.727	*.718*	−.410	−.342	−.587
NI	Intention	.720	.285	.588	.690	.573	.775	.672	−.480	−.313	*.812*	.986	.721
L	Loyalty	.612	.225	.561	.635	.455	.682	.579	−.392	−.249	.764	*.738*	.718
O	Overall	.756	.573	.677	.533	.326	.659	.581	−.696	−.497	.650	.617	–

The four emotion scales follow a two-dimensional arousal-valence model, but are often collapsed into positive and negative emotions (e.g., Doria et al. 2013). In the current analysis, the four basic emotion scales are used instead.

According to the theoretic background of AttrakDiff and meCUE and partially confirmed by the correlations in Table 9.3 (at least not disproved by the correlations), the sixteen scales are collapsed into four quality aspects:

Pragmatic Quality AttrakDiff Mini: Pragmatic Quality
                    meCUE: Effectiveness and Efficiency
Hedonic Quality AttrakDiff Mini: Hedonic Quality and Beauty
                    meCUE: Aesthetics, Status, and Commitment
Emotions meCUE: emotion scales (positive/negative valence crossed with high/low arousal)
Acceptance AttrakDiff Mini: Goodness
                    meCUE: Intention of Use, Loyalty, and Overall Rating

In the following, multivariate analyses are used to determine the strength of relationships between the experimental conditions, covariates, and interaction parameters and the subjective quality aspects. The use of multivariate statistics protects against alpha inflation that would occur if individual analyses for all sixteen scales were performed (Bortz 1999).

Before the relationship between UI design properties and user satisfaction can be analyzed, possible alternative influences have to be removed from the equation first. This is approached based on the covariates that were added to the experimental design.

Relationships between Covariates and Perceived Usability
To test the influence of the covariates on the subjective rankings, a single MANOVA with age, gender, OS, and browser as independent variables and all sixteen AttrakDiff Mini and meCUE scales as dependent variables was performed. The results are given in Table 9.4.

Because the covariate analysis reveals several strong relationships, and to rule out alternative explanations, the statistical analysis of the influence of UI design properties on the perceived usability will be performed on the residuals of the covariate analysis. Prior to this, different aspects of the UI design need to be operationalized based on the interaction logs.

Interaction Parameters
In order to quantify the influence of design parameters beyond the experimental conditions on the perceived usability, a set of four interaction parameters were derived from the individual action sequences of the participants.

logUserSteps The sum of all user steps during the experiment. Because this is extremely right-skewed, a natural logarithmic transformation was performed before statistical analysis.
logTaskTime The time in seconds between the loading of the first experimental screen (French restaurant task) and the completion of the last one (car

Table 9.3 Correlations between meCUE and AttrakDiff

AttrakDiff	meCUE											
	Pragmatic		Hedonic			Emotion				Acceptance		
	N	U	A	S	C	AP	DP	AN	DN	NI	L	O
Pragmatic Qu.	.576	.662	.352	.303	.140	.435	.397	−.577	−.407	.364	.379	.609
Hedonic Qu.	.628	.286	.779	.481	.246	.587	.477	−.499	−.372	.569	.496	.660
Beauty	.531	.277	.646	.390	.202	.452	.388	−.372	−.248	.457	.403	.550
Goodness	.636	.423	.586	.438	.208	.522	.431	−.547	−.408	.535	.488	.716

Table 9.4 Effect of covariables on AttrakDiff and meCUE scales

Covariate	η_G^2	Λ	df	p
Age	.119	.881	1	<.001**
Gender	.061	.939	1	.086[†]
OS	.047	.909	2	.248
Browser	.095	.905	1	.001**

Note **highly significant (p < .01); *significant (p < .05); [†]marginally significant (p < .10)

repair shop task). This was transformed to natural logarithmic scale as well.

propDevOr The proportion of device-oriented user steps.

propWait The proportion of time spent waiting for search results to appear during the experimental stage (dependent on responsivity of the external Shutterstock server).

Correlations between these four interaction parameters are given in Table 9.5.

Manipulation Check

The effects of the experimental manipulation on the four interaction parameters are given in Table 9.6. The three experimental conditions led to strong differences for propDevOr and logUserSteps. The logTaskTime differs less and propWait only marginally.

Table 9.5 Intercorrelations of the interaction parameters

	1 steps	2 time	3 devOr
1 logUserSteps	–		
2 logTaskTime	.823	–	
3 propDevOr	.667	.495	–
4 propWait	.462	.247	.260

Table 9.6 Effect of the three experimental conditions on the chosen interaction parameters (ANOVA results)

Interaction parameter	η_G^2	$F_{2,397}$	p
logUserSteps	.116	26.1	<0.001**
logTaskTime	.029	6.0	.003**
propDevOr	.302	85.7	<0.001**
propWait	.012	2.3	.097[†]

Note **highly significant (p < .01); *significant (p < .05); [†]marginally significant (p < .10)

Relationships between Interaction Parameters and Perceived Usability

To counteract the interrelation between the experimental conditions and the number of user steps, the cases were weighted based on their number of steps. For reasons of simplicity, the weighting algorithm treats both the general distribution of logUserSteps and the marginal distributions depending on the experimental conditions as Gaussian distributions. The weight of each data point is then computed as quotient of its probability density under the general and its marginal distribution. As a result, participants with long action sequences were weighted higher in the drag-from-overview condition and lower in the two other conditions (and vice versa).

Because the weighting affects the degrees of freedom of parametric statistical procedures like (M)ANOVA, bootstrapped permutation tests were used to determine statistical significance in the following analyses (Davison and Hinkley 1997). The remaining relationships between experimental condition and interaction parameters are given in Table 9.7.

The three experimental conditions only have a small influence on the subjective ratings, results of weighted MANOVAs on the residuals are given in Table 9.8. In an AUE scenario, it is of most interest whether this small difference could in principle be predicted based on the simulated interactions of the AUE tool. How do these relate to the subjective measures?

The influence of the interaction parameters on the subjective usability was analyzed based on the residuals of the covariate analysis above (Table 9.4). Individual weighted MANOVAs were computed for each of the four aspects. The results are given in Table 9.9.

Table 9.7 Weighting check: Statistical separation of the conditions and the logUserSteps and logTaskTime parameters. ANOVA with bootstrap permutation test (2000 runs)

Interaction parameter	η_G^2	F	p
logUserSteps	.005	.9	.438
logTaskTime	.008	1.5	.281
propDevOr	.173	41.6	<.001**
propWait	.011	2.1	.158

Note **highly significant (p < .01); *significant (p < .05); [†]marginally significant (p < .10)

Table 9.8 Influence of the experimental conditions on the subjective ratings. MANOVA with bootstrap permutation test (2000 runs)

Interaction parameter	η_G^2	Λ	p
Pragmatic quality	.007	.987	.596
Hedonic quality	.023	.955	.083[†]
Emotions	.019	.963	.102
Acceptance	.019	.962	.101

Note **highly significant (p < .01); *significant (p < .05); [†]marginally significant (p < .10)

Table 9.9 Influence of interaction parameters on perceived usability. MANOVA with bootstrap permutation test (2000 runs)

Factor	Pragmatic Qu.			Hedonic Qu.			Emotions			Acceptance		
	η_G^2	Λ	p	η_G^2	Λ	p	η_G^2	Λ	p	η_G^2	Λ	p
logUserSteps	.006	.994	.516	.015	.985	.385	.014	.986	.316	.022	.978	.093†
logTaskTime	.019	.981	.083†	.011	.989	.568	.008	.992	.608	.023	.976	.078†
propDevOr	.009	.991	.377	.012	.988	.491	.004	.996	.868	.030	.970	.027*
propWait	.033	.967	.007**	.012	.988	.484	.038	.962	.012*	.022	.978	.104

Note **highly significant (p < .01); *significant (p < .05); † marginally significant (p < .10)

Subsequent individual weighted ANOVAs were computed for the four scales within the acceptance aspect. The results are given in Table 9.10. The scale with the strongest relationship to device-orientation is the overall rating of meCUE, but it is only marginally significant.

Prediction of the Overall Acceptance

Based on both the theoretical background (Möller 2010; Thüring and Mahlke 2007) and the results above, the overall acceptance rating is chosen as operationalization of user satisfaction that should be predicted by an automated usability evaluation system. While the waiting time had substantial impact on the usability ratings in the experiment, this parameter would not be accessible to an AUE tool. The predictions are therefore based on logUserSteps and propDevOr, only.[4]

A linear model with only logUserSteps as independent variable explains 1.5% of the variance of the acceptance ratings. Adding propDevOr to the linear mode yields 3.0%, which is a significant increase ($F_{1,397} = 6.4, p = .012$). Adding the interaction does not yield substantial improvements.

9.1.3   Discussion

The strong relationships between perceived quality and the covariates, namely with age, replicate earlier findings (e.g., Wechsung 2014, see Table 9.4). The significantly different ratings depending on the web browser used are most likely due to browser-specific graphical renderings of the UI elements.

Regarding the actual research question, i.e., the impact of goal relevance (operationalized as propDevOr) on user satisfaction, the first unexpected finding is that measures of *pragmatic* quality (i.e., subjective efficiency and effectiveness) are *not* substantially influenced by the propDevOr parameter. While counterintuitive, this result is still consistent with literature as objective measures of effectiveness and efficiency are usually only weakly related to user satisfaction (Hornbæk and Law 2007). Interestingly, this is lesser the case for *subjective* measures of efficiency like perceived task completion time (Backhaus and Trapp 2015).

While the overall *acceptance* shows a statistically significant correlation with propDevOr, acceptance can only be explained by this UI property to a small degree ($\eta_G^2 = .03$, Table 9.9). This is again consistent with the results of the meta-analysis by Hornbæk and Law (2007) who found stronger correlations between objective measures and preference than between objective measures and satisfaction. One of the reasons for these findings is that usability ratings are very volatile and susceptible

---

[4]This may seem contradictory to the approach in Sect. 5.6 where system response time was a parameter of the MASP-MeMo-CogTool prediction. The difference here is that the wait time in the current section is a fluctuating parameter that depends on an external system that most probably would be unavailable at early design stages. The system response times that go into the model in Sect. 5.6 however are constant values like the time between user click and visual button state change which depend only on the UI code and can therefore be used for KLM-based time predictions.

Table 9.10 Influence of interaction parameters on acceptance ratings. ANOVA with bootstrap permutation test (2000 runs)

Factor	Goodness			Intent. of use			Loyalty			Overall rating		
	η^2_G	F	p	η^2_G	F	p	η^2_G	F	p	η^2_G	F	p
logUserSteps	.003	1.1	.334	.001	0.6	.468	.000	0.0	.854	.002	0.8	.405
logTaskTime	.016	6.5	.019*	.000	0.1	.790	.001	0.4	.583	.001	0.5	.481
propDevOr	.004	1.5	.258	.002	1.0	.354	.000	0.1	.842	.008	3.4	.078†
propWait	.021	8.4	.008**	.010	3.9	.064†	.004	1.6	.237	.009	3.7	.059†

Note **highly significant (p < .01); *significant (p < .05); † marginally significant (p < .10)

to manipulations (e.g., "what is beautiful is usable", Tractinsky et al. 2000; Raita and Oulasvirta 2011; Hamborg et al. 2014).

Finally, the strong relationships between the proportion of time spent waiting for search results (propWait) and pragmatic quality and emotions should not be accounted to the user striving for efficiency, as the relation between propWait and logTaskTime was much weaker (and both are related, i.e., propWait can account for the relationship to logTaskTime). Instead, this factor is connected to the system forcing the user to do nothing and leaves no option for further interaction. This leads to negative emotions and also negative attitudes towards the system because it deprives the user's need of *self-efficacy* ('competence' in the terms of Deci and Ryan 2000, see also Bandura, 1996). It is nevertheless interesting that while the feeling of competence is considered a *hedonic* aspect of usability (e.g., Hassenzahl 2007, p. 10), only the ratings of *pragmatic* quality are affected in this study. This should be mainly due to competence not being represented well in the item sets of AttrakDiff Mini and meCUE.

9.2 Conclusion

The current chapter highlights the limitations of automated usability evaluation in general. As Ivory and Hearst (2001, see quote on p. 23) have stated before, AUE cannot capture user preference. It is nevertheless interesting that a small but significant impact of goal relevance on user satisfaction could be established.

This result concludes the empirical work presented here to analyze the viability of UI meta-information as predictor of usability. The following final chapter will review the progress made towards that goal, discuss the validity and applied relevance of the findings and finish with conclusions and future work.

Chapter 10
General Discussion and Conclusion

In this chapter[1]:

- Recap the scientific contributions from the previous chapters.
- Discussion on the background of the research questions stated in the introduction.
- Concluding remarks and outlook.

The main research question of this work has been initially stated as how UI meta-information as created by the MBUID process can be used for automated usability evaluation (Sect. 1.4). The notion of MBUID meta-information has been narrowed down to the CAMELEON-conformant runtime models of the MASP at the end of Pt. I. The answers provided in following parts will now be reviewed and discussed.

10.1 Overview of the Contributions

Usability can be divided into three quality aspects: efficiency, effectiveness, and satisfaction (ISO 9241-11 1998). How these can be predicted on the basis of information contained in the model-based application framework MASP on the different abstraction layers FUI, AUI, and task model (CTT) is given below.

Prediction of Efficiency
The efficiency of a UI can be operationalized as the time needed to complete a specified task. The task completion time (TCT) can be predicted using the KLM (Card et al. 1983) mainly as sum of motor time (P and K operators), mental time (M operators), and system response time (SR operators).

---

[1]Parts of this chapter have already been published in Halbrügge et al. (2016).

© Springer International Publishing AG 2018
M. Halbrügge, *Predicting User Performance and Errors*, T-Labs Series
in Telecommunication Services, DOI 10.1007/978-3-319-60369-8_10

Previous AUE systems (e.g., CogTool; John et al. 2004) rely mainly on information on the FUI level (e.g., UI element placement) for predictions of TCT. As shown in Chap. 5, the incorporation of additional AUI information leads to an improved M operator placement that would otherwise need human intervention. Together with new KLM rules for system response time handling inspired by the DBDR strategy (Gray 2000, see Sect. 5.1), other FUI information (navigation to different page, button toggle action; see also Quade 2015) leads to improved SR operator placement. In total, the AUE system presented in that chapter provides much better predictions across several studies than the previous state-of-the-art (i.e., CogTool; John et al. 2004).

Prediction of Effectiveness

The effectiveness of a UI design can be measured based on the number of errors[2] that users make on their path towards a goal. In the error domain, no commonly accepted predictive model exists. Therefore, a link function from UI design to user errors had to be established, first.

This was approached based on the activation-based MFG theory (Altmann and Trafton 2002) and the concept of goal relevance (Ament et al. 2013). While a first analysis confirmed the importance of goal relevance, new data revealed that another design property needs to be taken into account as well: whether a UI element is *obligatory* or not. The empirical error rate was especially high for UI elements that had both low goal relevance (i.e., they were device-oriented) and that were not obligatory (see Table 6.6). Both goal relevance and task necessity (i.e., whether using a UI element is obligatory) can be derived from the MASP models (FUI, AUI, and CTT level), they are therefore promising UI properties for automated evaluation. Nevertheless, predictions on their basis should not only be based on data, they should also be psychologically plausible.

Two cognitive processes were therefore proposed as mediators between the UI properties and user errors: *task priming* is received by subgoals that correspond to task-oriented UI elements (high goal relevance) and *visual priming* increases the activation of subgoals that are corresponding to the current visual focus of attention. These processes are modulating goal-directed behavior that is following two distinct strategies: While the next subgoal can be retrieved, a *memory-based strategy* completes the current subgoal and tries to retrieve the next one. Once this fails, a second strategy performs *visual search* for UI elements that potentially correspond to uncompleted subgoals. An implementation of the priming processes and the two strategies in the cognitive architecture ACT-R (Anderson et al. 2004) confirmed that these can account for the empirical results, i.e., they are possible theoretical explanations of the error pattern. The existence of a visual cue-seeking strategy was additionally confirmed with eye-tracking.

A third factor that mediates between UI properties and error rates can be derived from knowledge bases (ontologies) that are external to the information contained in

---

[2] As errors often lead to corrective behavior (which takes additional time), this operationalization is not independent from the efficiency measure of task completion time. It nevertheless represents a sufficiently different aspect of usability.

the MASP models. The psychological process of *concept priming* captures the effect that UI elements that relate to highly familiar concepts (e.g., 'German' as opposed to 'lactose intolerance') are less prone to omissions than UI elements that do not. This effect is overall weaker than the combined effect of goal relevance and task necessity (see Table 7.3), but it provides an approach which is complementary to the other ones and also links to, e.g., the learnability of a system.

On the software engineering side, priming effects could be captured efficiently based on the LTMC system (Schultheis et al. 2006). An integrated MASP-MeMo-LTMC simulation could predict which UI elements of a new system were especially prone to errors in a dedicated validation study.

Prediction of User Satisfaction
User satisfaction is commonly measured using questionnaires. The subjective ratings of users are usually only weakly connected to objectively measured efficiency and effectiveness parameters (Hornbæk and Law 2007).

Of the UI properties introduced above, a significant relationship between the amount of device-oriented steps (or 'tool-task ratio'; Kirschenbaum et al. 1996) and system acceptance ratings could be found in a dedicated online study, but the effect is not very strong (3% explained variance).

10.2 General Discussion

The viability of the approach depends mainly on three things: the validity of the user model (and thereby, predictions), the applied relevance of the automated predictions, and the costs and benefits of the AUE approach compared to user testing.

10.2.1 Validity of the User Models

The validity of the models has already been discussed before, in particular in Sects. 5.8 and 8.5. In order to keep the general discussion focused, only the main points will be given here. The validity of the models is important for the applicability of an AUE system because it is a necessary (but not sufficient) prerequisite of the applicability of such a system. An AUE tool can only be useful if its predictions are valid.

In case of the integrated AUE solutions presented in this work, validity was first ensured by basing the user models on psychological theory instead of limited data, only. Thereby, the results of many more studies than the one presented here were (indirectly) incorporated into the models. This should increase their generalizability.

Second, newly proposed theoretical processes (task priming and visual cue seeking in Chap. 6, concept priming in Chap. 7) were approached empirically using several disjoint dependent variables (e.g., completion time, error rate, gaze behavior) to ensure their validity.

Finally, the models were tested on new applications or in new contexts to analyze not only the reproducibility of the effects found before, but also the models' generalization to new domains.

With respect to this, a major limiting factor was the small number of available MASP applications which restricted the experimentally observable user behavior mainly to tag-based search and list handling. Especially in the error domain, this has the disadvantage that some frequent sources of error, e.g., data entry (Greene and Tamborello 2015), could not be studied. Such errors are therefore not covered by the integrated system.

Another limitation in terms of validity stems from the general approach taken in this work. By basing the AUE on the development time models of the UI designers, the evaluation will hardly elicit misconceptions that the designers had during the process (e.g., bad wording or violation of established user workflows). The validity of the evaluation is bound to the validity of the information provided by the MASP. This problem may be alleviated in the future by the incorporation of external knowledge bases (see Sect. 7.1).

10.2.2   Applicability and Practical Relevance of the Predictions

Even if the validity of the AUE predictions is given, they may still be lacking practical relevance, e.g., if they are hard to obtain, hard to interpret, or do not yield information that is actually important for the users of an AUE tool.

In this domain, the integration of the user models with the MASP provides fast and easy results. More importantly, these results are available from early development stages on and could be *automatically* created alongside the artifacts at the end of each iteration cycle like weekly integration builds. Thereby, usability issues could be identified and fixed earlier and also with less costs as the costs of making changes to a design increases dramatically with the progress of the development (Nielsen 1993). The possibility to automate the evaluation also leads to *scalability* that is important for maintaining the usability of multi-target applications ("plasticity"; Coutaz and Calvary 2012).

The most complicated parts of the AUE process are the definition of the tasks in a computer readable manner (Sect. 8.2; see also Quade 2015) and the final interpretation of the simulation results. Of these two, the task definition syntax follows well-established document standards (XML; Bray et al. 1998), it also needs to be done only once and can be reused afterwards. The interpretation of the results may be more complicated, e.g., because the simulation outputs many numerical values like a task completion time of 8.723 s or an error rate of .0109. The apparent precision of these values may lead to overinterpretation. Relative displays like the rank-based Table 8.2 should therefore be preferred.

The practical relevance of task completion time and error predictions depends on the application domain, but is generally high. Valid time predictions may save employers millions of dollars (Gray et al. 1993), and human error can have fatal consequences (Reason 1990, 2016). In case of software systems that directly target end users (e.g., home stereo control instead of power plant control) the practical relevance of time and error predictions decreases. The link to the general acceptance of a system is weak as psychological needs of the users other than (objectively measured) efficiency and effectiveness play a larger role, here (see Chap. 9). Time and error predictions are nevertheless relevant in domains like safety-critical systems (e.g., machine control, finance), or when time is an important cost factor (e.g., enterprise software used in the workplace[3]). At first sight, this may pose a large restriction on the applicability of the approach, but enterprise software in fact accounted for about 75% of the worldwide software market in 2013[4] (300 out of 407 billion USD; Gartner 2014a, b).

Finally, the practical relevance is again limited by the system being bound to the MASP which is a very specific framework with few applications. The rather formal approach taken in this work ensures that the integrated system developed here should be adaptable to other model-based interaction systems as long as these follow the CAMELEON (Calvary et al. 2003) reference, though.

10.2.3   Costs and Benefits

The usefulness of the integrated system can be assessed at least anecdotally by comparing the time and money spent on the validation of the integrated system in Experiment 5 (Sect. 8.3) to the costs of the simulation. Both approaches share the initial task of planning the evaluation (three days). Running the validation experiment with 30 participants took six days, manually annotating the videos took another five days, and the statistical analysis another two days. The work was evenly shared between a researcher and a student worker, which leads to a conservatively estimated daily rate of 150€. The participants of the study were paid a total of 300€. Neglecting additional costs for equipment and room rent, this sums up to a total of 2700€ for the experiment, compared to 450€ for the simulation. Furthermore, simulating 100 users took two days on a standard consumer laptop, which is much faster than the eleven days of empirical data collection and video annotation.

---

[3]Example: In one observational study, office workers in different departments of a company spent most of their time using e-mail software (Peres 2005). According to Peres, e-mail handling was not only done rather inefficiently, the employees also lacked motivation to learn more efficient strategies as they considered e-mail being not that relevant (compared to, e.g., efficient handling of an integrated software development environment in case of software engineers). From the employer's perspective, this opens the possibility to reduce costs if e-mail software is designed for user efficiency.

[4]For reference: In that year, worldwide mobile app store revenue totaled 26 billion USD (Gartner 2013a) and video games without consoles 49 billion USD (Gartner 2013b).

When it comes to evaluating the plasticity (Coutaz and Calvary 2012) of adaptable UIs, the *scalability* of the empirical and the simulation approach becomes of highest importance. In the empirical case, money and time costs for conducting experiments and annotating videos multiply with the number of UI adaptations that need to be covered. The simulation, on the other hand, only needs more computational processing time for each new version of the FUI, making it possible to evaluate the usability of arbitrary numbers of different UIs at stable costs as long as the AUI and CTT models remain unchanged.

Finally, the UI of the health assistant needed to receive some polishing before the user tests could start which led to extra costs and time delay. The automated system on the other hand does not get distracted by broken images or skewed layouts that quickly grab the participants' attention during user studies. The last point is especially important during early design stages when no presentable UI is available.

10.3 Conclusion

UI meta-information that is provided by model-based interaction systems like the MASP is actually useful to provide improved predictions of the usability of applications that are developed using such interaction systems.

Due to this meta-information being computer-processable, usability predictions based on it can be created by automated systems that integrate model-based UI development with cognitive user modeling. This is especially useful during early design stages, when tests with real users are not yet possible, and during the development of multi-target applications with many target-specific versions of the UI, where classical user tests of all UI versions would be extremely costly and time-consuming.

The scope of the automated usability predictions spans both the efficiency (task completion time) and effectiveness (proneness to errors) of an application given a set of previously specified tasks. User satisfaction can not be covered to an extent that is practically relevant ($\eta_G^2 = .03$).

The development of an error prediction system has led to the elicitation of UI properties that affect human error that had not been researched before and has resulted in the formulation of a new theoretical model of human error that has been validated in several domains. This serves as a good example how solving an applied problem (here: using MBUID information for AUE) can lead to advances in psychological theory (here: better understanding of sequential control and human error) as well.

Future directions should include the integration of the user simulation into the development lifecycle, the integration of world knowledge from an external knowledge base into the MASP-MeMo-LTMC system, and a necessity and sensitivity analysis as performed for the initial error model (Sect. 6.4) should be performed for the full error model as well.

References

Agresti A (2014) Categorical data analysis. Wiley, New Jersey

Altmann EM, Trafton JG (2002) Memory for goals: an activation-based model. Cognit Sci 26(1):39–83. doi:10.1207/s15516709cog2601_2

Altmann EM, Trafton JG, Hambrick DZ (2014) Momentary interruptions can derail the train of thought. J Exp Psychol Gener 143(1):215–226. doi:10.1037/a0030986

Ament MG (2011a) Frankenstein and human error: device-oriented steps are more problematic than task-oriented ones. In: CHI'11: extended abstracts on human factors in computing systems. ACM, New York, NY, pp 905–910. doi:10.1145/1979742.1979514

Ament MG (2011b) The role of goal relevance in the occurrence of systematic slip errors in routine procedural tasks (Unpublished doctoral dissertation). UCL (University College London)

Ament MG, Cox AL, Blandford A, Brumby D (2010) Working memory load affects device-specific but not task-specific error rates. In: Ohlsson S, Catrambone R (eds) Proceedings of the 32nd annual conference of the cognitive science society. Portland, OR, pp 91–96

Ament MG, Cox AL, Blandford A, Brumby DP (2013) Making a task difficult: evidence that device-oriented steps are effortful and error-prone. J Exp Psychol Appl 19(3):195. doi:10.1037/a0034397

Anderson JR (2005) Human symbol manipulation within an integrated cognitive architecture. Cognit Sci 29(3):313–341. doi:10.1207/s15516709cog0000_22

Anderson JR (2007) How can the human mind occur in the physical universe?. Oxford University Press, Oxford, UK

Anderson JR, Bothell D, Byrne MD, Douglass S, Lebiere C, Qin Y (2004) An integrated theory of the mind. Psychol Rev 111(4):1036–1060. doi:10.1037/0033-295X.111.4.1036

Anderson JR, Bower GH (2014) Human associated memory. Psychology press, New York

Anderson JR, Lebiere C (1998) The atomic components of thought. Lawrence Erlbaum Associates, Mahwah

Anderson JR, Reder LM (1999) The fan effect: new results and new theories. J Exp Psychol Gener 128(2):186–197. doi:10.1037/0096-3445.128.2.186

Anderson JR, Zhang Q, Borst JP, Walsh MM (2016) The measurement of processing stages: extension of sternberg's method. Psychol Rev. doi:10.1037/rev0000030

Baber C, Stanton NA (1996) Human error identification techniques applied to public technology: predictions compared with observed use. Appl Ergon 27:119–131. doi:10.1016/0003-6870(95)00067-4

Backhaus N, Trapp AK (2015) Das ging ja flott! Zeitwahrnehmung im Usability- und UX-Testing. In: Wienrich C, Zander TO, Gramann K (eds) 11. Technische Universität Berlin, Berlin, Berliner Werkstatt Mensch-Maschine-Systeme, pp 61–65

Bandura A (1996) Self-efficacy: the exercise of control. Freeman, New York

Bates D, Maechler M, Bolker B, Walker S (2013) lme4: linear mixed-effects models using eigen and s4 [Computer software manual]. (R package version 1.0-5)

Bevan N (2009) Usability. In: Liu L, Özsu MT (eds) Encyclopedia of database systems. Boston, MA, Springer US, pp 3247–3251. doi:10.1007/978-0-387-39940-9_441

Blandford A, Green TR, Furniss D, Makri S (2008) Evaluating system utility and conceptual fit using cassm. Int J Hum Comput Stud 66(6):393–409. doi:10.1016/j.ijhcs.2007.11.005

Blumendorf M, Feuerstack S, Albayrak S (2008) Multimodal user interfaces for smart environments: the multi-access service platform. In: AVI'08: Proceedings of the working conference on advanced visual interfaces. ACM, New York, NY, USA, pp 478–479. doi:10.1145/1385569.1385665

Blumendorf M, Lehmann G, Albayrak S (2010) Bridging models and systems at runtime to build adaptive user interfaces. In: Proceedings of the 2nd ACM SIGCHI symposium on engineering interactive computing systems. ACM, New York, NY, pp 9–18. doi:10.1145/1822018.1822022

Blumendorf M, Lehmann G, Feuerstack S, Albayrak S (2008) Executable models for human-computer interaction. In: Graham TCN, Palanque P (eds) DSV-IS 2008: 15th international workshop on design, specification, and verification of interactive systems. Springer, Berlin, pp 238–251. doi:10.1007/978-3-540-70569-7_22

Bolton ML, Bass EJ, Siminiceanu RI (2012) Generating phenotypical erroneous human behavior to evaluate human-automation interaction using model checking. Int J Hum Comput Stud 70(11):888–906. doi:10.1016/j.ijhcs.2012.05.010

Borst JP, Ghuman AS, Anderson JR (2016) Tracking cognitive processing stages with MEG: a spatio-temporal model of associative recognition in the brain. NeuroImage 141:416–430. doi:10.1016/j.neuroimage.2016.08.002

Bortz J (1999) Statistik für sozialwissenschaftler, 5th edn. Springer, Berlin

Botvinick MM, Bylsma LM (2005) Distraction and action slips in an everyday task: evidence for a dynamic representation of task context. Psychon Bull Rev 12(6):1011–1017

Bray T, Paoli J, Sperberg-McQueen CM, Maler E, Yergeau F (1998) Extensible markup language (XML) (World Wide Web Consortium Recommendation No. REC-xml-19980210). http://www.w3.org/TR/1998/REC-xml-19980210

Brysbaert M, Buchmeier M, Conrad M, Jacobs AM, Bölte J, Böhl A (2011) The word frequency effect: a review of recent developments and implications for the choice of frequency estimates in German. Exp Psychol 58(5):412–424. doi:10.1027/1618-3169/a000123

Butterworth R, Blandford A, Duke D (2000) Demonstrating the cognitive plausibility of interactive system specifications. Formal Asp Comput 12(4):237–259. doi:10.1007/s001650070021

Byrne MD (2013) Computational cognitive modeling of interactive performance. In: Lee JD, Kirlik A (eds) The oxford handbook of cognitive engineering. Oxford University Press, pp 415–423

Byrne MD, Bovair S (1997) A working memory model of a common procedural error. Cognit Sci 21(1):31–61. doi:10.1207/s15516709cog2101_2

Byrne MD, Davis EM (2006) Task structure and postcompletion error in the execution of a routine procedure. Hum Factors J Hum Factors Ergon Soc 48(4):627–638. doi:10.1518/001872006779166398

Calvary G, Coutaz J, Thevenin D, Limbourg Q, Bouillon L, Vanderdonckt J (2003) A unifying reference framework for multi-target user interfaces. Int Comput 15(3):289–308. doi:10.1016/S0953-5438(03)00010-9

Card SK, Moran TP, Newell A (1983) The psychology of human-computer interaction. Erlbaum Associates, New Jersey

Clerckx T, Luyten K, Coninx K (2004) Dynamo-aid: a design process and a runtime architecture for dynamic model-based user interface development. In: Engineering human computer interaction and interactive systems, pp 77–95. doi:10.1007/11431879_5

Cooper R, Shallice T (2000) Contention scheduling and the control of routine activities. Cognit Neuropsychol 17(4):297–338. doi:10.1080/026432900380427

Coutaz J, Calvary G (2012) HCI and software engineering for user interface plasticity. In Jacko JA (ed) Human-computer interaction handbook: fundamentals, evolving technologies, and emerging applications, 3rd ed. CRC Press, pp 1195–1220

Cox AL, Young RM (2000) Device-oriented and task-oriented exploratory learning of interactive devices. In: Taatgen NA, Aasman J (eds) Proceedings of the third international conference on cognitive modeling. Universal Press, Veenendaal, NL, pp 70–77

Davies HTO, Crombie IK, Tavakoli M (1998) When can odds ratios mislead? BMJ Br Med J 316(7136):989–991. doi:10.1136/bmj.316.7136.989

Davison AC, Hinkley DV (1997) Bootstrap methods and their application. Cambridge University Press, New York

Deci EL, Ryan RM (2000) The "what" and "why" of goal pursuits: human needs and the self-determination of behavior. Psychol Inq 11(4):227–268. doi:10.1207/S15327965PLI1104_01

Doria L, Minge M, Riedel L, Kraft M (2013) User-centred evaluation of lower-limb orthoses: a new approach. Biomed Eng/Biomedizinische Technik 58(Suppl. 1): doi:10.1515/bmt-2013-4232

Ecker UK, Lewandowsky S, Oberauer K, Chee AEH (2010) The components of working memory updating: an experimental decomposition and individual differences. J Exp Psychol Learn Mem Cognit 36(1):170. doi:10.1037/a0017891

Engelbrecht K-P (2013) Estimating spoken dialog system quality with user models. Springer, Berlin

Engelbrecht K-P, Kruppa M, Möller S, Quade M (2008) MeMo workbench for semiautomated usability testing. In: Interspeech, pp 1662–1665

Fitts PM (1954) The information capacity of the human motor system in controlling the amplitude of movement. J Exp Psychol 47(6):381–391. doi:10.1037/h0055392

Frohlich D (1997) Direct manipulation and other lessons. In: Helander M, Landauer TK, Prabhu P (eds) Handbook of human-computer interaction. Elsevier Science BV, Amsterdam, pp 463–488

Fu W-T, Pirolli P (2007) SNIF-ACT: a cognitive model of user navigation on the world wide web. Hum Comput Int 22:355–412. doi:10.1080/07370020701638806

Gartner (2013a) Gartner says mobile app stores will see annual downloads reach 102 billion in 2013. Press release. http://www.gartner.com/newsroom/id/2592315. Accessed 08 Aug 2016

Gartner (2013b) Gartner says worldwide video game market to total $93 billion in 2013. Press release. http://www.gartner.com/newsroom/id/2614915. Accessed 08 Aug 2016

Gartner (2014a) Gartner says worldwide it spending on pace to grow 2.1 percent in 2014. Press release. http://www.gartner.com/newsroom/id/2783517. Accessed 08 Aug 2016

Gartner (2014b) Gartner says worldwide software market grew 4.8 percent in 2013. Press release. http://www.gartner.com/newsroom/id/2696317. Accessed 08 Aug 2016

Gluck KA, Stanley CT, Moore LR, Reitter D, Halbrügge M (2010) Exploration for understanding in cognitive modeling. J Artif Gener Intell 2(2):88–107. doi:10.2478/v10229-011-0011-7

Goodfellow I, Courville A, Bengio Y (2015) Deep learning. http://goodfeli.github.io/dlbook/ (Draft Version 2015-12-3)

Gould JD, Lewis C (1985) Designing for usability: key principles and what designers think. Commun ACM 28(3):300–311. doi:10.1145/3166.3170

Gray WD (2000) The nature and processing of errors in interactive behavior. Cognit Sci 24(2):205–248. doi:10.1016/S0364-0213(00)00022-7

Gray WD (2008) Cognitive architectures: choreographing the dance of mental operations with the task environment. Hum Factors J Hum Factors Ergon Soc 50(3):497–505. doi:10.1518/001872008X312224

Gray WD, Fu W-T (2004) Soft constraints in interactive behavior: the case of ignoring perfect knowledge in-the-world for imperfect knowledge in-the-head. Cognit Sci 28(3):359–382. doi:10.1016/j.cogsci.2003.12.001

Gray WD, John BE, Atwood ME (1993) Project Ernestine: validating a GOMS analysis for predicting and explaining real-world task performance. Hum Comput Int 8(3):237–309. doi:10.1207/s15327051hci0803_3

Greene KK, Tamborello F (2015) Password entry errors: Memory or motor? In: Taatgen NA, van Vugt MK, Borst JP, Mehlhorn K (eds) Proceedings of the 13th international conference on cognitive modeling. University of Groningen, Groningen, The Netherlands, pp 226–231

Guse D (2016) TheFragebogen. http://thefragebogen.de/. Accessed 25 July 2016

Halbrügge M (2007) Evaluating cognitive models and architectures. In Kaminka GA, Burghart CR (eds) Evaluating architectures for intelligence. papers from the 2007 AAAI workshop. AAAI Press, Menlo Park, California, pp 27–31. http://www.aaai.org/Papers/Workshops/2007/WS-07-04/WS07-04-007.pdf

Halbrügge M (2013) ACT-CV: Bridging the gap between cognitive models and the outer world. In: Brandenburg E, Doria L, Gross A, Günzlera T, Smieszek H (eds) Grundlagen und Anwendungen der Mensch-Maschine-Interaktion– 10. Berliner Werkstatt Mensch-Maschine- Systeme. Universitätsverlag der TU Berlin, Berlin, pp 205–210. doi:10.14279/depositonce-3802

Halbrügge M (2015a) Automatic online analysis of eye-tracking data for dynamic HTML-based user interfaces. In: Wienrich C, Zander TO, Gramann K (eds) 11. Berliner Werkstatt Mensch-Maschine-Systeme. Technische Universität Berlin, Berlin, pp 322–324. doi:10.14279/depositonce-4887

Halbrügge M (2015b) Fast-time user simulation for dynamic HTML-based interfaces. In: Taatgen NA, van Vugt MK, Borst JP, Mehlhorn K (eds) Proceedings of the 13th international conference on cognitive modeling. University of Groningen, Groningen, the Netherlands, pp 51–52

Halbrügge M (2016a) Rethinking the keystroke-level model from an embodied cognition perspective. In: Barkowsky T, Llansola ZF, Schultheis H, van de Ven J (eds) KogWis 2016: 13th biannual conference of the german cognitive science society, pp 51–54

Halbrügge M (2016b) Towards the evaluation of cognitive models using anytime intelligence tests. In: Reitter D, Ritter FE (eds) Proceedings of the 14th international conference on cognitive modeling. Penn State, University Park, PA, pp 261–263. http://acs.ist.psu.edu/iccm2016/proceedings/halbruegge2016iccmB.pdf

Halbrügge M, Engelbrecht K-P (2014) An activation-based model of execution delays of specific task steps. Cognit Process 15:S107–S110

Halbrügge M, Engelbrecht K-P (2015) Können Nutzer im usability-labor zwischen interface-varianten unterscheiden? Zwei Fallbeispiele aus dem Smart Home. In: Wienrich C, Zander TO, Gramann K (eds) 11. Berliner Werkstatt Mensch-Maschine-Systeme. Technische Universität Berlin, Berlin, pp 27–32. doi:10.14279/depositonce-4887

Halbrügge M, Quade M, Engelbrecht K-P (2015a) How can cognitive modeling benefit from ontologies? Evidence from the HCI domain. In: Bieger J, Goertzel B, Potapov A (eds) Proceedings of AGI 2015, vol 9205. Springer, Berlin, pp 261–271. doi:10.1007/978-3-319-21365-1_27

Halbrügge M, Quade M, Engelbrecht K-P (2015b) A predictive model of human error based on user interface development models and a cognitive architecture. In: Taatgen NA, van Vugt MK, Borst JP, Mehlhorn K (eds) Proceedings of the 13th international conference on cognitive modeling. University of Groningen, Groningen, the Netherlands, pp 238–243

Halbrügge M, Quade M, Engelbrecht K-P (2016) Cognitive strategies in HCI and their implications on user error. In: Papafragou A, Grodner D, Mirman D, Trueswell JC (eds) Proceedings of the 38th annual meeting of the cognitive science society. Cognitive Science Society, Austin, TX, pp 2549–2554

Halbrügge M, Quade M, Engelbrecht K-P, Möller S, Albayrak S (2016) Predicting user error for ambient systems by integrating model-based UI development and cognitive modeling. In: Ubi-Comp'16: The 2016 ACM international joint conference on pervasive and ubiquitous computing. ACM, New York, NY. doi:10.1145/2971648.2971667

Halbrügge M, Russwinkel N (2016) The sum of two models: how a composite model explains unexpected user behavior in a dual-task scenario. In: Reitter D, Ritter FE (eds) Proceedings of the 14th international conference on cognitive modeling. Penn State, University Park, PA, pp 137–143. http://acs.ist.psu.edu/iccm2016/proceedings/halbruegge2016iccm.pdf

Halbrügge M, Schultheis H (2016) Modeling kitchen knowledge with LTMC. In: Barkowsky T, Llansola ZF, Schultheis H, van de Ven J (eds) KogWis 2016: 13th biannual conference of the german cognitive science society, pp 83–86

Hamborg K-C, Hülsmann J, Kaspar K (2014) The interplay of usability and aesthetics: more evidence for the "what is usable is beautiful" notion. Adv Hum Comput Int. doi:10.1155/2014/946239

Hanisch T (2014) A compound semantic analyzing module for the automated usability evaluation framework maspmemo (Bachelor's Thesis). Freie Universität Berlin, Berlin, Germany

Hassenzahl M (2007) The hedonic/pragmatic model of user experience. In: Law E, Vermeeren A, Hassenzahl M, Blythe M (eds) Towards a UX manifesto, pp 10–14

Hassenzahl M, Beu A, Burmester M (2001) Engineering joy. IEEE Softw 18(1):70. doi:10.1109/52.903170

Hassenzahl M, Monk A (2010) The inference of perceived usability from beauty. Hum Comput Int 25(3):235–260. doi:10.1080/07370024.2010.500139

Hassenzahl M, Wiklund-Engblom A, Bengs A, Hägglund S, Diefenbach S (2015) Experience-oriented and product-oriented evaluation: psychological need fulfillment, positive affect, and product perception. Int J Hum Comput Int 31(8):530–544. doi:10.1080/10447318.2015.1064664

Hiatt LM, Trafton JG (2015) An activation-based model of routine sequence errors. In: Taatgen NA, van Vugt MK, Borst JP, Mehlhorn K (eds) Proceedings of the 13th international conference on cognitive modeling. University of Groningen, Groningen, the Netherlands, pp 244–249

Hiltz K, Back J, Blandford A (2010) The roles of conceptual device models and user goals in avoiding device initialization errors. Int Comput 22(5):363–374. doi:10.1016/j.intcom.2010.01.001

Hollnagel E (1993) The phenotype of erroneous actions. Int J Man Mach Stud 39(1):1–32. doi:10.1006/imms.1993.1051

Hollnagel E (1998) Cognitive reliability and error analysis method (CREAM). Elsevier, Oxford, UK

Hornbæk K, Law EL-C (2007) Meta-analysis of correlations among usability measures. In: CHI'07: Proceedings of the SIGCHI conference on human factors in computing systems, pp 617–626. doi:10.1145/1240624.1240722

ISO 9241–11, (1998) Ergonomic requirements for office work with visual display terminals (VDTs)– Part 11: Guidance on usability. International Organization for Standardization, Geneva, Switzerland

ISO 9241–210, (2010) Ergonomics of human-system interaction–Part 210: Human-centred design for interactive systems. International Organization for Standardization, Geneva, Switzerland

Ivory MY, Hearst MA (2001) The state of the art in automating usability evaluation of user interfaces. ACM Comput Surv (CSUR) 33(4):470–516. doi:10.1145/503112.503114

Jameson A, Mahr A, Kruppa M, Rieger A, Schleicher R (2007) Looking for unexpected consequences of interface design decisions: the MeMo workbench. In: Winckler M, Johnson H, Palanque P (eds) Proceedings of the 6th international conference on task models and diagrams for user interface design, TAMODIA'07. Springer, Berlin. doi:10.1007/978-3-540-77222-4_24

John BE (1990) Extensions of GOMS analyses to expert performance requiring perception of dynamic visual and auditory information. In: CHI'90: Proceedings of the SIGCHI conference on human factors in computing systems. New York, pp 107–116. doi:10.1145/97243.97262

John BE, Jastrzembski TS (2010) Exploration of costs and benefits of predictive human performance modeling for design. In: Proceedings of the 10th international conference on cognitive modeling. Philadelphia, PA, pp 115–120

John BE, Kieras DE (1996a) The GOMS family of user interface analysis techniques: comparison and contrast. ACM Trans Comput Hum Int (TOCHI) 3(4):320–351. doi:10.1145/235833.236054

John BE, Kieras DE (1996b) Using GOMS for user interface design and evaluation: which technique? ACM Trans Comput Hum Int (TOCHI) 3(4):287–319. doi:10.1145/235833.236050

John BE, Prevas K, Salvucci DD, Koedinger K (2004) Predictive human performance modeling made easy. In: CHI'04: Proceedings of the SIGCHI conference on human factors in computing systems ACM Press, New York, USA, pp 455–462. doi:10.1145/985692.985750

Kieras DE (1999) A guide to GOMS model usability evaluation using GOMSL and GLEAN3 (Technical Report). University of Michigan

Kieras DE, Santoro TP (2004) Computational GOMS modeling of a complex team task: lessons learned. In: CHI'04: Proceedings of the SIGCHI conference on human factors in computing systems, pp 97–104. doi:10.1145/985692.985705

Kieras DE, Wood SD, Abotel K, Hornof A (1995) GLEAN: a computer-based tool for rapid GOMS model usability evaluation of user interface designs. In: UIST'95: Proceedings of the 8th annual ACM symposium on user interface and software technology, pp 91–100. doi:10.1145/215585. 215700

Kirschenbaum SS, Gray WD, Ehret BD, Miller SL (1996) When using the tool interferes with doing the task. In: CHI'96: conference companion on human factors in computing systems, pp 203–204. doi:10.1145/257089.257281

Kirwan B (1997a) Validation of human reliability assessment techniques: part 1–validation issues. Saf Sci 27(1):25–41. doi:10.1016/S0925-7535(97)00049-0

Kirwan B (1997b) Validation of human reliability assessment techniques: part 2–validation results. Saf Sci 27(1):43–75. doi:10.1016/S0925-7535(97)00050-7

Kirwan B, Ainsworth LK (1992) A guide to task analysis: the task analysis working group. CRC press

Langley P (2016) An architectural account of variation in problem solving and execution. In: Papafragou A, Grodner D, Mirman D, Trueswell JC (eds) Proceedings of the 38th annual meeting of the cognitive science society. Cognitive Science Society, Austin, TX, pp 2843–2844

Lehmann J, Isele R, Jakob M, Jentzsch A, Kontokostas D, Mendes PN, Bizer C (2015) DBpedia–a large-scale, multilingual knowledge base extracted from Wikipedia. Semant Web 6(2):167–195. doi:10.3233/SW-140134

Li SY, Blandford A, Cairns P, Young RM (2008) The effect of interruptions on postcompletion and other procedural errors: an account based on the activation-based goal memory model. J Exp Psychol Appl 14(4):314. doi:10.1037/a0014397

Limbourg Q, Vanderdonckt J, Michotte B, Bouillon L, López-Jaquero V (2005) USIXML: a language supporting multi-path development of user interfaces. In: Bastide R, Palanque P, Roth J (eds) Engineering human computer interaction and interactive systems, vol 3425, pp 200–220. Springer, Berlin. doi:10.1007/11431879_12

Lord FM, Novick MR, Birnbaum A (1968) Statistical theories of mental test scores. Addison-Wesley, Reading, MA

Mayhew DJ (1999) The usability engineering lifecycle. In: CHI'99 extended abstracts on human factors in computing systems, pp 147–148. doi:10.1145/632716.632805

McGlashan S et al (eds) (2004) Voice extensible markup language (VoiceXML) version 2.0 (Technical Report). W3C Recommendation. https://www.w3.org/TR/voicexml20/. Accessed 08 Aug 2016

Meixner G, Paternò F, Vanderdonckt J (2011) Past, present, and future of model-based user interface development. i-com, 10(3), 2–11. doi:10.1524/icom.2011.0026

Miller J, Mukerji J (2001) Model driven architecture (MDA) (Technical Report No. ormsc/2001-07-01). Object management group, architecture board ORMSC. http://www.omg.org/cgi-bin/doc?ormsc/01-07-01.pdf. Accessed 08 Aug 2016

Minge M, Riedel L (2013) meCUE–Ein modularer Fragebogen zur Erfassung des Nutzungserlebens. In: Boll S, Maass S, Malaka R (eds) Mensch und Computer 2013: interaktive Vielfalt, pp 89–98. München

Möller S (2010) Quality Engineering: Qualität kommunikationstechnischer Systeme. Springer, Berlin

Möller S, Englert R, Engelbrecht K-P, Hafner VV, Jameson A, Oulasvirta A, Reithinger N (2006) MeMo: towards automatic usability evaluation of spoken dialogue services by user error simulations. Proceedings of the 9th international conference on spoken language processing (Interspeech 2006 - ICSLP). ISCA, Pittsburgh, PA, pp 1786–1789

Mori G, Paternò F, Santoro C (2002) CTTE: support for developing and analyzing task models for interactive system design. IEEE Trans Softw Eng 28(8):797–813. doi:10.1109/TSE.2002. 1027801

Mori G, Paternò F, Santoro C (2004) Design and development of multidevice user interfaces through multiple logical descriptions. IEEE TransSoftw Eng 30(8):507–520. doi:10.1109/TSE.2004.40

Naderi B, Wechsung I, Möller S (2015) Effect of being observed on the reliability of responses in crowdsourcing micro-task platforms. In: Seventh international workshop on quality of multimedia experience (QoMEX), pp 1–2. doi:10.1109/QoMEX.2015.7148091

Newell A, Simon HA (1972) Human problem solving. Prentice-Hall, Englewood Cliffs, NJ

Nielsen J (1993) Usability engineering. Academic Press, San Diego, CA

Nielsen J, Landauer TK (1993) A mathematical model of the finding of usability problems. In: Proceedings of the INTERACT'93 and CHI'93 conference on human factors in computing systems. ACM, New York, NY, pp 206–213. doi:10.1145/169059.169166

Norman DA (1981) Categorization of action slips. Psychol Rev 88(1):1. doi:10.1037/0033-295X.88.1.1

Norman DA (1988) The psychology of everyday things. Basic books, New York, NY

Norman DA, Shallice T (1986) Attention to action: willed and automatic control of behavior. In: Davidson RJ, Schwartz GE, Shapiro D (eds) Consciousness and self-regulation: advances in theory and research. Plenum Press, New York, NY, pp 1–18. (Revised reprint of Norman & Shallice, 1980)

Page L, Brin S, Motwani R, Winograd T (1999) The PageRank citation ranking: bringing order to the web. (Technical Report No. 1999-66). Stanford InfoLab. http://ilpubs.stanford.edu:8090/422/

Palanque P, Basnyat S (2004) Task patterns for taking into account in an efficient and systematic way both standard and erroneous user behaviours. In: Human error, safety and systems development. Springer, pp 109–130. doi:10.1007/1-4020-8153-7_8

Paternò F (2003) ConcurTaskTrees: an engineered notation for task models. In: Diaper D, Stanton N (eds) The handbook of task analysis for human-computer interaction. Lawrence Erlbaum Associates, Mahwah, NJ, pp 483–501

Paternò F, Santoro C (2002) Preventing user errors by systematic analysis of deviations from the system task model. Int J Hum Comput Stud 56(2):225–245. doi:10.1006/ijhc.2001.0523

Patton EW, Gray WD, John BE (2012) Automated CPM-GOMS modeling from human data. Proceedings of the human factors and ergonomics society annual meeting 56:1005–1009. doi:10.1177/1071181312561210

Peres SC (2005) Software use in the workplace: a study of efficiency (Unpublished doctoral dissertation). Rice Univercity, Houston, TX

Pinheiro J, Bates D, DebRoy S, Sarkar D, Core Team R (2013) nlme: linear and nonlinear mixed effects models [Computer software manual]. (R package version 3.1-113)

Pirolli P (1997) Computational models of information scent-following in a very large browsable text collection. In: CHI'97: Proceedings of the ACM SIGCHI conference on human factors in computing systems, pp 3–10. doi:10.1145/258549.258558

Plumbaum T, Narr S, Eryilmaz E, Hopfgartner F, Klein-Ellinghaus F, Reese A, Albayrak S (2014) Providing multilingual access to health-related content. In: Lovis C, Seroussi B, Hasman A, Pape-Haugaard L, Saka O, Andersen SK (eds) eHealth—for continuity of care: Proceedings of MIE2014. IOS Press, Amsterdam, NL, pp 393–397. doi:10.3233/978-1-61499-432-9-393

Quade M (2015) Automation in model-based usability evaluation of adaptive user interfaces by simulating user interaction (Doctoral dissertation, Fakultät IV, Technische Universität Berlin). doi:10.14279/depositonce-4918

Quade M, Halbrügge M, Engelbrecht K-P, Albayrak S, Möller S (2014) Predicting task execution times by deriving enhanced cognitive models from user interface development models. In: Proceedings of the 2014 ACM SIGCHI symposium on engineering interactive computing systems. ACM, New York, NY, USA, pp 139–148. doi:10.1145/2607023.2607033

Core Team R (2014) R: a language and environment for statistical computing [Computer software manual]. Vienna, Austria. http://www.R-project.org

Raggett D, Le Hors A, Jacobs I (eds) (1999) HTML 4.01 specification (Technical Report). W3C Recommendation. https://www.w3.org/TR/html401/. Accessed 08 Aug 2016

Raita E, Oulasvirta A (2011) Too good to be bad: favorable product expectations boost subjective usability ratings. Int Comput 23(4):363–371. doi:10.1016/j.intcom.2011.04.002

Raskin J (1997) Looking for a humane interface: will computers ever become easy to use? Commun ACM 40(2):98–101. doi:10.1145/253671.253737

Rasmussen J (1983) Skills, rules, and knowledge; signals, signs, and symbols, and other distinctions in human performance models. IEEE Trans Syst Man Cybern 13:257–266. doi:10.1109/TSMC. 1983.6313160

Rasmussen J (1986) Information processing and human-machine interaction: an approach to cognitive engineering. North Holland, New York

Ratwani RM, Trafton JG (2011) A real-time eye tracking system for predicting and preventing postcompletion errors. Hum Comput Int 26(3):205–245. doi:10.1080/07370024.2011.601692

Reason J (1990) Human error. Cambridge University Press, New York, NY

Reason J (2016) Organizational accidents revisited. CRC Press, Boca Raton, FL

Roberts S, Pashler H (2000) How persuasive is a good fit? a comment on theory testing. Psychol Rev 107(2):358–367. doi:10.1037/0033-295X.107.2.358

Ruh N, Cooper RP, Mareschal D (2010) Action selection in complex routinized sequential behaviors. J Exp Psychol Hum Percept Perform 36(4):955. doi:10.1037/a0017608

Rukšēenas R, Curzon P, Blandford A, Back J (2014) Combining human error verification and timing analysis: a case study on an infusion pump. Form Asp Comput 26:1033–1076. doi:10. 1007/s00165-013-0288-1

Russwinkel N, Urbas L, Thüring M (2011) Predicting temporal errors in complex task environments: a computational and experimental approach. Cognit Syst Res 12(3):336–354. doi:10. 1016/j.cogsys.2010.09.003

Salvucci DD (2006) Modeling driver behavior in a cognitive architecture. Hum Factors 48(2):362–380. doi:10.1518/001872006777724417

Salvucci DD (2009) Rapid prototyping and evaluation of in-vehicle interfaces. ACM Trans Comput Hum Int (TOCHI) 16(2):9

Salvucci DD (2010) On reconstruction of task context after interruption. In: CHI'10: Proceedings of the SIGCHI conference on human factors in computing systems, pp 89–92. doi:10.1145/1753326. 1753341

Salvucci DD (2014) Endowing a cognitive architecture with world knowledge. In: Bello P, Guarini M, McShane M, Scassellati B (eds) Proceedings of the 36th annual meeting of the cognitive science society, pp 1353–1358

Salvucci DD, Goldberg JH (2000) Identifying fixations and saccades in eye-tracking protocols. In: Proceedings of the 2000 symposium on eye tracking research & applications, pp 71–78. doi:10. 1145/355017.355028

Salvucci DD, Taatgen NA (2008) Threaded cognition: an integrated theory of concurrent multitasking. Psychol Rev 115(1):101–130. doi:10.1037/0033-295X.115.1.101

Sanchez M, Barrero I, Villalobos J, Deridder D (2008) An execution platform for extensible runtime models. In: 3rd workshop on Models@run.time at models'08, pp 107–116

Schaffer S, Schleicher R, Möller S (2015) Modeling input modality choice in mobile graphical and speech interfaces. Int J Hum Comput Stud 75:21–34. doi:10.1016/j.ijhcs.2014.11.004

Schmidt S, Engelbrecht K-P, Schulz M, Meister M, Stubbe J, Töppel M, Möller S (2010) Identification of interactivity sequences in interactions with spoken dialog systems. In: PQS 2010: 3rd international workshop on perceptual quality of systems, pp 109–114

Schultheis H (2009) Computational and explanatory power of cognitive architectures: The case of act-r. In: Howes A, Peebles D, Cooper RP (eds) Proceedings of the 9th international conference on cognitive modeling. Manchester, UK

Schultheis H, Barkowsky T, Bertel S (2006) LTM C—an improved long-term memory for cognitive architectures. In: Proceedings of the seventh international conference on cognitive modeling, pp 274–279

Schultheis H, Lile S, Barkowsky T (2007) Extending ACT-R's memory capabilities. In: Proceedings of EuroCogSci'07: the European cognitive science conference. Lawrence Erlbaum Associates, pp 758–763

Schulz M (2016) Simulation des Interaktionsverhaltens von Senioren bei der Benutzung von mobilen Endgeräten (Doctoral dissertation, Fakultät IV, Technische Universität Berlin). doi:10. 14279/depositonce-4991

Schwartz MF, Montgomery MW, Buxbaum LJ, Lee SS, Carew TG, Coslett HB, Mayer N (1998) Naturalistic action impairment in closed head injury. Neuropsychology 12(1):13–28. doi:10.1037/ 0894-4105.12.1.13

Simon HA, Newell A (1971) Human problem solving: the state of the theory in 1970. Am Psychol 26(2):145. doi:10.1037/h0030806

Singleton WT (1973) Theoretical approaches to human error. Ergonomics 16(6):727–737. doi:10. 1080/00140137308924563

Sottet J-S, Calvary G, Coutaz J, Favre J-M (2008) A model-driven engineering approach for the usability of plastic user interfaces. In: Gulliksen J, Harning MB, Palanque P, van der Veer GC, Wesson J (eds) Engineering interactive systems, vol 4940. Springer, Berlin, pp 140–157. doi:10. 1007/978-3-540-92698-6_9

Statistisches Bundesamt (2016) Ausstattung privater Haushalte mit Informations- und Kommunikationstechnik im Zeitvergleich. https://www.destatis.de/DE/ZahlenFakten/ GesellschaftStaat/EinkommenKonsumLebensbedingungen/AusstattungGebrauchsguetern/ Tabellen/A_Infotechnik_D_LWR.html. Accessed 08 Aug 2016

Stewart TC, West RL (2010) Testing for equivalence: a methodology for computational cognitive modelling. J Artif Gener Intell 2(2):69–87. doi:10.2478/v10229-011-0010-8

Taatgen NA, Van Rijn H, Anderson JR (2007) An integrated theory of prospective time interval estimation: the role of cognition, attention, and learning. Psychol Rev 114(3):577. doi:10.1037/ 0033-295X.114.3.577

Tamborello FP, Trafton JG (2015) Action selection and human error in routine procedures. Proceedings of the human factors and ergonomics society annual meeting 59:667–671. doi:10.1177/ 1541931215591145

Teo L, John BE (2008) Towards a tool for predicting goal-directed exploratory behavior. Proceedings of the human factors and ergonomics society annual meeting 52:950–954. doi:10.1177/ 154193120805201311

Thüring M, Mahlke S (2007) Usability, aesthetics and emotions in human-technology interaction. Int J Psychol 42(4):253–264. doi:10.1080/00207590701396674

Tognazzini B (1992) Tog on interface. Addison-Wesley, Reading, MA

Tractinsky N, Katz A, Ikar D (2000) What is beautiful is usable. Int Comput 13(2):127–145. doi:10. 1016/S0953-5438(00)00031-X

Trafton JG, Altmann EM, Ratwani RM (2011) A memory for goals model of sequence errors. Cognit Syst Res 12:134–143. doi:10.1016/j.cogsys.2010.07.010

Ulich E, Rauterberg M, Moll T, Greutmann T, Strohm O (1991) Task orientation and useroriented dialog design. Int J Hum Comput Int 3(2):117–144. doi:10.1080/10447319109526001

Vanderdonckt J (2005) A MDA-compliant environment for developing user interfaces of information systems. In: Pastor O, Falcão e Cunha J (eds) CAiSE 2005: 17th international conference on advanced information systems engineering. Springer, Berlin, pp 16–31. doi:10.1007/11431855_ 2

Veksler VD, Myers CW, Gluck KA (2015) Model flexibility analysis. Psychol Rev 122:755–769. doi:10.1037/a0039657

Vera AH, John BE, Remington R, Matessa M, Freed MA (2005) Automating human-performance modeling at the millisecond level. Hum Comput Int 20(3):225–265. doi:10.1207/ s15327051hci2003_1

Wechsung I (2014) An evaluation framework for multimodal interaction. Springer, Berlin. doi:10. 1007/978-3-319-03810-0

Westermann R (2000) Wissenschaftstheorie und Experimentalmethodik: Ein Lehrbuch zur psychologischen Methodenlehre. Hogrefe, Göttingen

Wickens CD, Hollands JG, Banbury S, Parasuraman R (2015) Engineering psychology & human performance. Psychology Press

Wilcox RR (2005) Comparing medians: an overview plus new results on dealing with heavy-tailed distributions. J Exp Educ 73(3):249–263. doi:10.3200/JEXE.73.3.249-263

Wilson M (2002) Six views of embodied cognition. Psychon Bull Rev 9(4):625–636. doi:10.3758/BF03196322

Wittenburg P, Brugman H, Russel A, Klassmann A, Sloetjes H (2006) ELAN: a professional framework for multimodality research. In: Proceedings of lrec, vol 2006

Index

© Springer International Publishing AG 2018

147

M. Halbrügge, *Predicting User Performance and Errors*, T-Labs Series
in Telecommunication Services, DOI 10.1007/978-3-319-60369-8

Printed in the United States
By Bookmasters